Geoengineering of the Climate System

ISSUES IN ENVIRONMENTAL SCIENCE AND TECHNOLOGY

How to obtain future titles on publication:

A subscription is available for this series. This will bring delivery of each new volume immediately on publication and also provide you with online access to each title via the Internet. For further information visit http://www.rsc.org/issues or write to the address below.

For further information please contact:
Sales and Customer Care, Royal Society of Chemistry, Thomas Graham House, Science Park, Milton Road, Cambridge, CB4 0WF, UK
Telephone: +44 (0)1223 432360, Fax: +44 (0)1223 426017, Email: booksales@rsc.org
Visit our website at www.rsc.org/books

ISSUES IN ENVIRONMENTAL SCIENCE AND TECHNOLOGY

EDITORS: R.E. HESTER AND R.M. HARRISON

38
Geoengineering of the Climate System

Issues in Environmental Science and Technology No. 38

Print ISBN: 978-1-84973-953-5
PDF eISBN: 978-1-78262-122-5
ISSN: 1350-7583

A catalogue record for this book is available from the British Library

Published by The Royal Society of Chemistry,
Thomas Graham House, Science Park, Milton Road,
Cambridge CB4 0WF, UK

Registered Charity Number 207890

For further information see our web site at www.rsc.org

Printed in the United Kingdom by CPI Group (UK) Ltd, Croydon, CR0 4YY, UK

Preface

Evidence that the climate is warming is overwhelming. Not only does this include land and sea surface temperature records, but also other indicators such as the coverage of Arctic sea ice. Few serious scientists doubt that anthropogenic activities, and specifically the emissions of greenhouse gases, are responsible for the major part of the observed warming. Projections of future global temperatures under a "business-as–usual" scenario vary between models but indicate that future global warming will probably well exceed the 2 °C global average commonly regarded as the upper tolerable limit, and serious consequences in terms of temperature extremes, storminess, droughts and floods can be expected unless action is taken. The broad consensus in the scientific community is that the much preferable form of action would be to effect a dramatic reduction in emissions of greenhouse gases, and particularly of carbon dioxide so as to stabilise, and if possible, reduce current atmospheric concentrations. To do this solely through a reduction in emissions is not only challenging but might prove politically unacceptable, and consequently other actions have been proposed. Such actions, which come under the general term "geoengineering", include: interventions to stimulate natural processes of carbon dioxide removal (such as sequestration by terrestrial vegetation or within the oceans); chemical scrubbing from the atmosphere, changing planetary albedo; as well as more extreme measures such as using mirrors in space to intercept solar radiation before it reaches the earth; or creating a reflective aerosol layer in the stratosphere as a means of reducing solar input to the Earth's surface. This volume examines the scientific and engineering aspects of such options, as well as considering some of the associated governance issues.

The first chapter by John Thornes and Francis Pope describes the phenomenon of global warming and poses the question of why do we need solutions to global warming? It also sets the scene by describing the major options for geoengineering and considering the arguments for and against geoengineering research. This is followed by a chapter in which Stuart

Issues in Environmental Science and Technology, 38
Geoengineering of the Climate System
Edited by R.E. Hester and R.M. Harrison
© The Royal Society of Chemistry 2014
Published by the Royal Society of Chemistry, www.rsc.org

Haszeldine and Vivian Scott describe methods of carbon dioxide reduction which offer long timescales of storage. These are both on land and within the ocean, and the considerable complexities associated with seemingly straightforward measures are highlighted. The following chapter by Timothy Lenton goes further into the issue of carbon dioxide removal, looking at the overall global potential if all possible carbon sinks were optimised. The magnitude of the many possible removal options is assessed and a final conclusion reached that carbon dioxide removal has the physical potential to help stabilise atmospheric CO_2 in the middle of this century, if combined with reductions in carbon dioxide emissions. A complementary approach is to scrub carbon dioxide from the atmosphere by chemical means and Klaus Lackner explains the principles by which this might be achieved by means of "artificial trees" and estimates the likely financial cost of doing so. While current costs would be a huge increment on energy prices, the economics might become more favourable in the future.

Alternatives to carbon dioxide removal include those that aim to reflect more of the incoming solar radiation back to space, hence increasing the Earth's albedo, or reflectivity. One possibility is to adopt the growth of crops with a higher potential for reflectivity than those currently grown. Taraka Davies-Barnard sets out the basic principles and demonstrates what could be achieved by such a policy and concludes that it could make a useful though not large contribution to climate change mitigation. A complementary approach over the oceans involves creating artificial clouds by spraying large volumes of seawater into the atmosphere as fine droplets. Stephen Salter, Thomas Stevenson and Andreas Tsiamis describe the engineering technologies that would be needed to implement such measures, assess their possible effectiveness and also consider the likely economic costs. Moving to greater altitudes, Alan Robock describes how injections of sulfur gases into the stratosphere could be used to create a reflective sulfate layer which would lead to reduced surface temperatures. More detailed analysis shows that insolation reduction by this mechanism could keep the global average temperature constant, but global average precipitation would reduce particularly in summer monsoon regions, temperature changes would not be uniform and sea ice would continue to melt. Governance issues relating to such measures are also discussed. A perhaps more extreme option is to interpose large reflective objects or particle clouds in space between the sun and planet Earth. Colin McInnes, Russell Bewick and Joan Pau Sanchez describe the physical principles behind such technology and looks at the feasibility of its achievement.

Geoengineering raises huge ethical and governance issues which relate not only to the implementation of the technology but also to the research conducted into its feasibility. Alan Robock touches on these issues in his chapter, and in the final chapter of the book, Richard Owen gives an in-depth analysis of the issues and highlights some of his own experience in relation to research on geoengineering.

Geoengineering is a highly topical subject and hence very appropriate for coverage by the Issues in Environmental Technology series. We have been very fortunate to commission articles from some of the world's leading experts in this field and we believe that this volume provides an authoritative and highly informative overview not only of how geoengineering might be achieved but also of the likely financial costs, consequences and disbenefits. This volume will prove of value to a very wide range of scientists and engineers dealing with global issues, as well as policy-makers, and students in a wide range of environmental science and engineering courses.

Ronald E. Hester
Roy M. Harrison

Contents

Issues in Environmental Science and Technology, 38
Geoengineering of the Climate System
Edited by R.E. Hester and R.M. Harrison
© The Royal Society of Chemistry 2014
Published by the Royal Society of Chemistry, www.rsc.org

The Global Potential for Carbon Dioxide Removal   52
Timothy M. Lenton

The Use of Artificial Trees   80
Klaus S. Lackner

Engineering Ideas for Brighter Clouds   **131**
Stephen H. Salter, Thomas Stevenson and Andreas Tsiamis

Stratospheric Aerosol Geoengineering   **162**
Alan Robock

Editors

Ronald E. Hester, BSc, DSc (London), PhD (Cornell), FRSC, CChem

Ronald E. Hester is now Emeritus Professor of Chemistry in the University of York. He was for short periods a research fellow in Cambridge and an assistant professor at Cornell before being appointed to a lectureship in chemistry in York in 1965. He was a full professor in York from 1983 to 2001. His more than 300 publications are mainly in the area of vibrational spectroscopy, latterly focusing on time-resolved studies of photoreaction intermediates and on biomolecular systems in solution. He is active in environmental chemistry and is a founder member and former chairman of the Environment Group of the Royal Society of Chemistry and editor of 'Industry and the Environment in Perspective' (RSC, 1983) and 'Understanding Our Environment' (RSC, 1986). As a member of the Council of the UK Science and Engineering Research Council and several of its sub-committees, panels and boards, he has been heavily involved in national science policy and administration. He was, from 1991 to 1993, a member of the UK Department of the Environment Advisory Committee on Hazardous Substances and from 1995 to 2000 was a member of the Publications and Information Board of the Royal Society of Chemistry.

Roy M. Harrison, BSc, PhD, DSc (Birmingham), FRSC, CChem, FRMetS, Hon MFPH, Hon FFOM, Hon MCIEH

Roy M. Harrison is Queen Elizabeth II Birmingham Centenary Professor of Environmental Health in the University of Birmingham. He was previously Lecturer in Environmental Sciences at the University of Lancaster and Reader and Director of the Institute of Aerosol Science at the University of Essex. His more than 400 publications are mainly in the field of environmental chemistry, although his current work includes studies of human health impacts of atmospheric pollutants as well as research into the chemistry of pollution phenomena. He is a past Chairman of the Environment Group of the Royal Society of Chemistry for whom he has edited 'Pollution: Causes, Effects and Control' (RSC, 1983;

Fourth Edition, 2001) and 'Understanding our Environment: An Introduction to Environmental Chemistry and Pollution' (RSC, Third Edition,
1999). He has a close interest in scientific and policy aspects of air pollution,
having been Chairman of the Department of Environment Quality of
Urban Air Review Group and the DETR Atmospheric Particles Expert Group.
He is currently a member of the DEFRA Air Quality Expert Group, the
Department of Health Committee on the Medical Effects of Air Pollutants,
and Committee on Toxicity.

List of Contributors

Russell Bewick, OHB System AG, Universitätsallee 27–29, Bremen 28359, Germany. E-mail: rusbewick@hotmail.com

Taraka Davies-Barnard, School of Geographical Sciences, University of Bristol, University Road, Bristol BS8 1SS, United Kingdom. E-mail: t.davies-barnard@bristol.ac.uk

Stuart Haszeldine, Grant Institute, The King's Buildings, West Mains Road, Edinburgh EH9 3JW, United Kingdom, Email: s.haszeldine@ed.ac.uk

Klaus S. Lackner, Columbia University, Lenfest Center for Sustainable Energy and Department of Earth and Environmental Engineering, New York, NY 10027, USA. Email: kl2010@columbia.edu

Tim M. Lenton, College of Life and Environmental Sciences, University of Exeter, Level 7, Laver Building, Exeter EX4 4QE, United Kingdom. Email: t.m.lenton@exeter.ac.uk

Colin R. McInnes, Department of Mechanical and Aerospace Engineering, University of Strathclyde, Glasgow G1 1XJ, United Kingdom. E-mail: colin.mcinnes@strath.ac.uk

Richard Owen, Business School, University of Exeter, Rennes Drive, Exeter, EX4 4PU, United Kingdom. E-mail: r.j.owen@exeter.ac.uk

Francis D. Pope, School of Geography, Earth and Environmental Sciences, University of Birmingham, Edgbaston, Birmingham B15 2TT, United Kingdom. E-mail: f.pope@bham.ac.uk

Alan Robock, Department of Environmental Sciences, School of Environmental and Biological Sciences, Rutgers University, 14 College Farm Road, New Brunswick, New Jersey 08901-8551, USA. E-mail: robock@envsci.rutgers.edu

Stephen H. Salter, Institute for Energy Systems, School of Engineering, University of Edinburgh, Mayfield Road, Edinburgh EH9 3JL, Scotland, United Kingdom. Email: S.Salter@ed.ac.uk

Joan Pau Sanchez, Department of Applied Mathematics, Universitat Politècnica de Catalunya, Calle Jordi Girona, Barcelona 08034, Spain. E-mail: jpau.sanchez@upc.edu

Vivian Scott, Grant Institute, The King's Buildings, West Mains Road, Edinburgh EH9 3JW, United Kingdom. E-mail: Vivian.scott@ed.ac.uk

Thomas Stevenson, Institute for Energy Systems, School of Engineering, University of Edinburgh, Mayfield Road, Edinburgh EH9 3JL, Scotland, United Kingdom. Email: Tom.Stevenson@ee.ed.ac.uk

John Thornes, School of Geography, Earth and Environmental Sciences, University of Birmingham, Edgbaston, Birmingham B15 2TT, United Kingdom. E-mail: j.e.thornes@bham.ac.uk

Andreas Tsiamis, Scottish Microelectronics Centre, University of Edinburgh, West Mains Road, Edinburgh EH9 3JF, Scotland, United Kingdom. Email: A.Tsiamis@ed.ac.uk

Why do we need Solutions to Global Warming?

JOHN E. THORNES* AND FRANCIS D. POPE

ABSTRACT

The atmosphere is the most valuable resource on the planet and as such every effort needs to be made to protect and manage it. Unfortunately the rise in greenhouse gases since the industrial revolution, and the intimately linked change in climate, is proving to be a most difficult environmental problem. Even though the strongest scientific evidence tells us that the anthropogenic release of greenhouse gases is responsible for climate change, there has been little success in emissions reduction. The reasons behind this failure are complex but the outcome is not; the regions of the Earth inhabited by humans are on average getting hotter and extreme weather is becoming more frequent. Since mitigation efforts against climate change are failing, the arguments for the possibility of geoengineering become louder. Geoengineering is a contentious issue which evokes strong reactions within all levels of society. Solar Radiation Management (SRM) technologies are more controversial than Carbon Dioxide Removal (CDR) technologies, since they do not solve the root cause of the problem, they do, however, potentially offer a more rapidly deployed solution. At present no geoengineering technology is fit for purpose or ready for deployment. However, geoengineering research is rapidly increasing with hundreds if not thousands of scientists and engineers working on the topic worldwide. As such, geoengineering research has now likely passed through its infancy, and conclusions are being reached about

---

*Corresponding author

Issues in Environmental Science and Technology, 38
Geoengineering of the Climate System
Edited by R.E. Hester and R.M. Harrison
© The Royal Society of Chemistry 2014
Published by the Royal Society of Chemistry, www.rsc.org

the efficacy, benefits and disadvantages of the different proposals. It seems increasingly likely that geoengineering technologies could be developed that will reduce climate change. These benefits need to be carefully weighed against the negative aspects. A true assessment of geoengineering cannot be achieved until we better understand the environmental, technological, economic and governance issues associated through its use.

On May 9, 2013 the daily mean concentration of atmospheric carbon dioxide levels passed 400 ppm at Mauna Loa according to independent measurements taken by both the National Oceanic and Atmospheric Administration (NOAA) and the Scripps Institution of Oceanography. NOAA had announced last year that its global cooperative air sampling network had detected 400 ppm for the first time over all its Arctic sites, just a prelude to what is now being detected over Mauna Loa. According to NOAA, locations throughout the Southern Hemisphere will follow over the next few years, as the increase in Northern Hemisphere levels is always a little ahead of the Southern Hemisphere, due to the fact that the majority of carbon dioxide producing behemoths are found in the Northern Hemisphere.[1]

1   Introduction – Life and the Evolution of the Earth's Atmosphere

Life and the Earth's current atmosphere are intimately linked. You can't have one without the other. Imagine what would happen to the atmosphere if life was wiped out by the gamma rays of a supernova or by a supervirus that killed every living cell on the planet. The Earth would slowly convert, over 100 million years or so, to a planet much like Venus.[2] It would be hotter than the Earth's atmosphere before life, as the sun was about 30% fainter then, than it is now. Thus the atmosphere, weather and climate that we enjoy today are completely dependent on the abundance of life. Lovelock[3,4] powerfully shows us, through the metaphor of Gaia, that the Earth carefully self-regulates the thin layers of land, ocean and atmosphere to provide a flourishing environment for life. However to achieve a lasting symbiosis of mutual benefit to both the host (Earth) and the invader (life) can we prevent an eventual 'Tragedy of the Commons'?[5] The human population continues to exploit and pollute the atmosphere and 'foul its own nest', for the pursuit of energy and growth, supposedly for the benefit of today's 7 billion citizens and the 9 billion citizens expected by 2050. As a result, the Earth's climate is changing and we have already seen a rise in the planet's surface temperature of 0.8 °C due to radiative forcing caused by greenhouse gas emissions and land-use changes. This global warming is predicted to raise global mean surface temperatures by up to 5 °C by the end of this century if emissions of greenhouse gases continue to rise in a 'business as usual' fashion. Global warming is set to double even if we cease to emit any further pollution, due to the slow release of energy already stored in the oceans. This additional energy available to the atmosphere has already led to an increase in extreme weather around the globe and agreement that a realistic limit of 2 °C could well be surpassed.

The Intergovernmental Panel on Climate Change (IPCC) was set up in 1988 by two United Nations organisations: the World Meteorological Organisation (WMO) and the United Nations Environment Programme (UNEP) to critically assess the scientific, technical and socio-economic consequences of climate change and to examine options for society to mitigate greenhouse gas emissions and adapt to changing weather and climate. The IPCC is in the process of submitting its fifth set of *Assessment Reports* (AR1 in 1990, AR2 in 1995, AR3 in 2001, AR4 in 2007 and AR5 in 2014, http://www.ipcc.ch/) to support the *United Nations Framework Convention on Climate Change (UNFCCC, https://unfccc.int/kyoto_protocol/items/2830.php)*, an international treaty that set up the Kyoto Protocol which became effective in 2005. In the first commitment period (2008–2012), the Kyoto Protocol sought to set binding targets for 37 industrial countries and 15 European Union (EU-15) countries, on four greenhouse gases: carbon dioxide (CO_2), methane (CH_4), nitrous oxide (N_2O), sulphur hexafluoride (SF_6) and two groups of ozone depleting gases: hydrofluorocarbons (HFCs) and perfluorocarbons (PFCs).

The binding targets were modest and even so the results have been disappointing. Progress towards a new agreement (2012–2020) has been unsatisfactory because of the impasse in limiting the growth of greenhouse gas emissions. Alternative approaches (Plan B) such as geoengineering and the United Nations initiative *Sustainable Energy for All (SE4ALL)* are therefore being seriously considered.[6]

The focus of this chapter is on the options to sustainably manage this problem to prevent the atmosphere being polluted to such an extent that changes to the climate will be irreversible and damaging. Time is running out for solutions to be found that can be implemented in a sustainable way.

Firstly, we will scrutinize the value of the services that the atmosphere provides for society and their sensitivity to change. Secondly, we will look at how the climate is changing due to anthropogenic activity and the impacts that it is having on examples such as extreme weather, sea level rise, melting glaciers and ice caps. Thirdly, we will define geoengineering, and fourthly, we will examine the broad arguments for and against geoengineering, and the likely success of geoengineering as an instrument to manage the atmosphere should the mitigation of greenhouse gases fail to deliver.

2 The Atmosphere – The Most Valuable Resource on the Planet

Today's atmosphere has evolved slowly over more than 4 billion years. Changes in the composition of the atmosphere to what it is today are directly attributable to the development of living micro-organisms. Deliberate and inadvertent interventions, by forms of life, into the composition and behaviour of the atmosphere are consequently not new. Animal life (including humans) has evolved to become totally dependent on the atmosphere. The word 'animal' comes from the Latin word *animalis*, meaning 'having breath'. Typically on average we humans breathe about 15 m^3 of air per day. Without the air that we breathe we would die within minutes. Yet we take the atmosphere totally for granted. It is not just the air that we breathe that is vital. The atmosphere provides us with a whole range of 'atmospheric services' that are more valuable than any other resource on the planet.[7] Table 1 lists 12 of these services that are key to all life on Earth.

Typically the atmosphere is portrayed in the media as a hazard with almost daily tragedies caused by floods, droughts, gales, tornadoes, typhoons/

Table 1 The twelve atmospheric services.[7]

Rank in Value	Atmospheric Services	Usage Trend	At Risk	Entity	Service Type
1	The air that we breathe	++	**	O_2, N_2 etc.	Provisioning
2	Protection from radiation, plasma and meteors	+	**	Density, ozone layer	Supporting
3	Natural global warming of 33 degrees Celsius	+	*****	CO_2, CH_4, N_2O, H_2O^{++}	Supporting
4	The cleaning capacity of the atmosphere and dispersion of air pollution	+	*	OH, wind, temperature	Regulating
5	The redistribution of water services	+	**	H_2O	Supporting
6	Direct use of the atmosphere for ecosystems and agriculture	+	*	CO_2, N_2, filtered solar	Provisioning & Supporting
7	Combustion of fuel	–		O_2	Provisioning
8	Direct use of the atmosphere for sound, communications and transport	+	*	Density, pressure	Supporting
9	Direct use of the atmosphere for power	++		Wind, wave	Provisioning
10	The extraction of atmospheric gases	+		O_2, N_2, Ar etc.	Provisioning
11	Atmospheric recreation and climate tourism	+	*	Sun, temperature, wind, snow	Cultural
12	Aesthetic, spiritual and sensual properties of the atmosphere, smell and taste	+		Sky, clouds, rainbows etc.	Cultural

NatCatSERVICE

Weather catastrophes worldwide 1980 – 2012

Overall and insured losses with trend

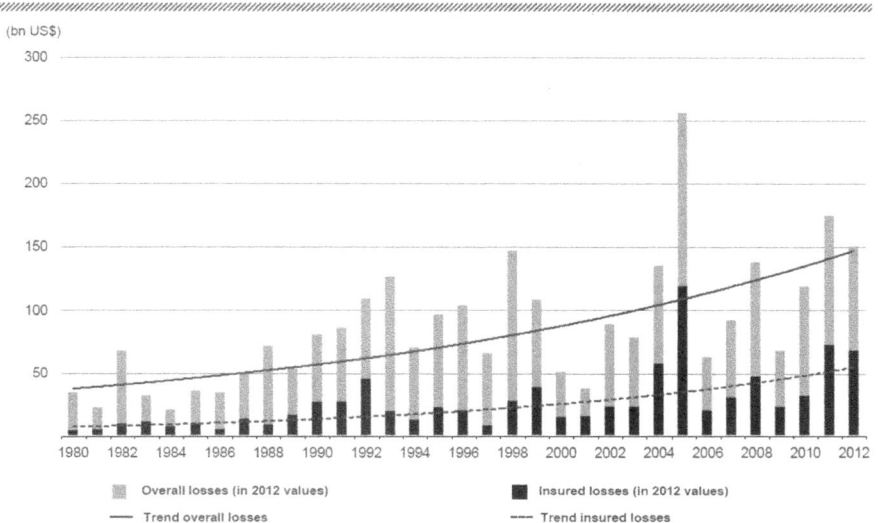

Figure 1   Overall and insured losses for weather catastrophes worldwide 1980–2012. (Data from Munich Re). (http://www.munichre.com/en/reinsurance/ business/non-life/georisks/natcatservice/default.aspx)

hurricanes, heat waves, snow and ice storms. The insurance group, Munich Re, compiles the best database of the worldwide number and costs of such hazards. During the period 1980–2012 they have estimated that there have been 18 200 weather catastrophes costing US$ 2.8 trillion (at 2012 prices) with the loss of 1 405 000 lives. They identify an upward trend that has seen a doubling in the annual number and cost of weather catastrophes since 1990 (see Figure 1). In 2012 there were more than 800 weather catastrophes logged at an estimated cost of US$ 150 billion.

Poor air quality is another vital issue that adds to the annual cost of breathing a polluted atmosphere. The World Health Organisation (WHO) considers clean air to be a basic requirement of human health and well-being. The European Commission has declared 2013 to be the 'Year of Air' and will take the opportunity to review current European air quality legislation. A recent estimate suggests that poor air quality is responsible for more than two million deaths worldwide each year:[8]

We estimate that in the present-day, anthropogenic changes to air pollutant concentrations since the preindustrial era are associated annually with 470 000 (95% confidence interval, 140 000 to 900 000) premature respiratory deaths related to ozone, and 2.1 (1.3 to 3.0) million CPD (cardiopulmonary *disease) and LC (lung cancer) deaths related to $PM_{2.5}$... We estimate here that 1500 premature respiratory deaths related to ozone and 2200 CPD and LC deaths related to $PM_{2.5}$ occur each year due to past climate change.*

The number of weather catastrophes is undoubtedly increasing due to climate change whereas the impact of climate change on air pollution is relatively small. Overall the impact of the atmosphere as a hazard to human life would be much worse if the atmosphere did not disperse air pollutants. In total, however, the atmosphere is worth orders of magnitude more as a resource than it costs as a hazard.[7]

The well mixed atmosphere is approximately 100 km deep, which is very thin in comparison to the size of the Earth. Indeed the troposphere, the lowest part of the earth's atmosphere is only about 8–12 km over the poles and 15–18 km deep over the equator and most of the life on the planet survives within the lowest 5 km which comprises half the atmosphere by weight. The effective atmosphere is therefore extremely thin, vulnerable and fragile and has been compared in size to the varnish on a globe. Hence the atmosphere is always taken for granted as it is effectively invisible and free.

3   The Greenhouse Effect and Global Warming

The natural greenhouse effect is responsible for keeping the global mean surface temperature 33 °C warmer than it would otherwise be. Without the atmosphere the mean temperature of the Earth would be −18 °C but with the atmosphere the mean global surface temperature is +15 °C. This natural greenhouse effect is caused by greenhouse gases in the atmosphere that are basically transparent to incoming solar radiation but trap and re-emit the Earth's thermal infrared radiation at certain wavelengths. This warms the atmosphere and the analogy of the workings of a greenhouse has been adopted for simplicity by policy makers (in reality the warming effect of a greenhouse depends on other factors too such as sheltering the air inside from the wind). Water vapour in the atmosphere is the most important natural greenhouse gas accounting for approximately 29.4 °C (89%) of the 33 °C. Carbon dioxide is only responsible for about 7.5% of the remaining 11% of natural warming.

On May 9, 2013 the daily mean concentration of atmospheric carbon dioxide levels passed 400 ppm at Mauna Loa, a background site in the middle of the Pacific Ocean, well away from industrial sources. Figures 2 and 3 show the steady upward rise of atmospheric CO_2 at Mauna Loa. There is no sign of a levelling off despite Kyoto, the global recession and other global attempts at mitigation. Annual CO_2 emissions from fossil fuel combustion and cement production were about 9.5 GtC (giga tons of carbon) in 2011, an increase of 54% over 1990 levels. From 1750 to 2011, cumulative global anthropogenic CO_2 emissions amount to approximately 545 GtC, of which 240 GtC have accumulated in the atmosphere, 155 GtC have been taken up

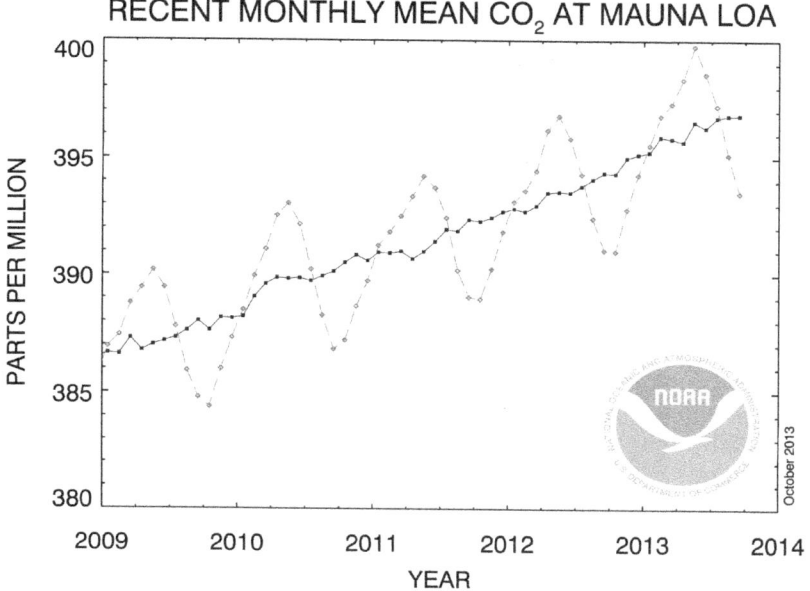

Figure 2   Recent monthly mean CO_2 at Mauna Loa.
(Data from the NOAA Earth System Research Laboratory, http://www.esrl.noaa.gov/gmd/ccgg/trends/).

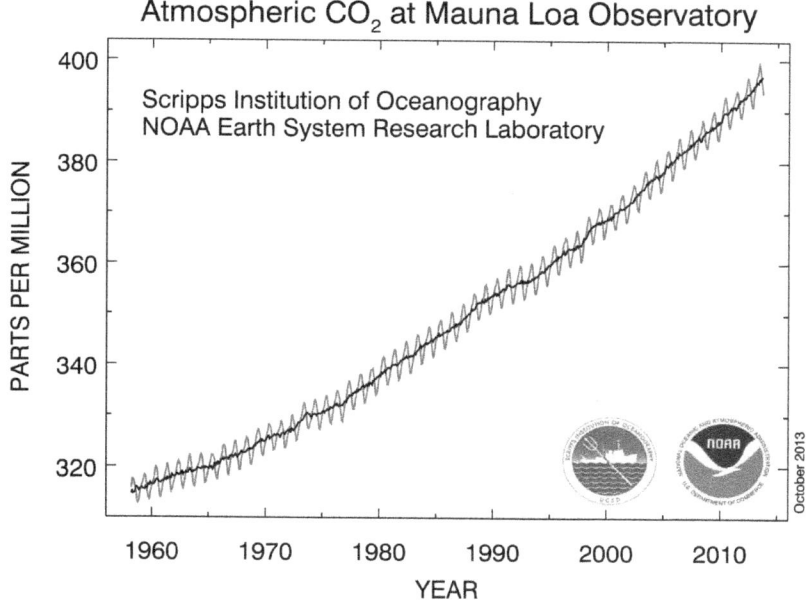

Figure 3   Atmospheric CO_2 at Mauna Loa observatory 1958–2013.
(Data from the Scripps Institution of Oceanography and NOAA Earth System Research Laboratory, http://www.esrl.noaa.gov/gmd/ccgg/trends/).

by the ocean and 150 GtC have accumulated in natural terrestrial ecosystems.

Global warming is the increase (0.8 °C so far) in the mean global surface temperature, above the 33 °C caused by the natural greenhouse effect, due to the human emitted greenhouse gases and also aerosol particles that directly absorb solar radiation. Carbon dioxide is responsible for nearly half of this increase. Figure 4 shows the latest estimate (for 2011) of radiative forcing caused by emissions of greenhouse gases and other drivers of change such as aerosols and black carbon. Radiative forcing of the climate drives climate change across the planet due to the uptake of additional energy into the climate system. The solar constant is on average 1365 Wm^{-2} and the additional total anthropogenic radiative forcing relative to 1750 is estimated by the IPCC (AR5) to be 2.29 (1.13 to 3.33) Wm^{-2}. This estimate is 43% higher than the estimate in AR4 (2005) due to the continued increase in greenhouse gases and an adjustment to give a weaker negative forcing caused by aerosols. There is not a simple linear relationship between radiative forcing and the global mean surface temperature increase as shown in Figure 5. Natural variability of the climate means that the global warming signal is superimposed on the noise of the natural greenhouse effect. However the IPCC (2013) state:[9]

> *Warming of the climate system is unequivocal and since the 1950s, many of the observed changes are unprecedented over decades to millennia. The atmosphere and ocean have warmed, the amounts of snow and ice have diminished, sea level has risen, and the concentrations of greenhouse gases have increased.*[9] *(p.2)*

What are the measurable impacts of global warming so far on the climate and the Earth's surface environment? Although the mean surface temperature rises have been observed across most of planet there is less certainty about precipitation.[9] There is some evidence for increased precipitation since 1901 over land areas in the northern hemisphere with greater intensities of precipitation also in North America and Europe. Over the oceans, however, more evidence is needed. Since 1950 more extreme weather and climate events have been observed. On the global scale the number of warm days and nights has increased, as have the number of heat waves. The number of cold days and nights has decreased despite a number of recent cold winters in the northern hemisphere.

> *Ocean warming dominates the increase in energy stored in the climate system, accounting for more than 90% of the energy accumulated between 1971 and 2010.*[9] *(p.6)*

The warming of the upper oceans is especially important and it has been estimated that of this 90%, it is likely that 60% is in the upper oceans (0–700 m) and 30% below 700 m. This energy will eventually be released into the atmosphere and add significantly to global warming this century.

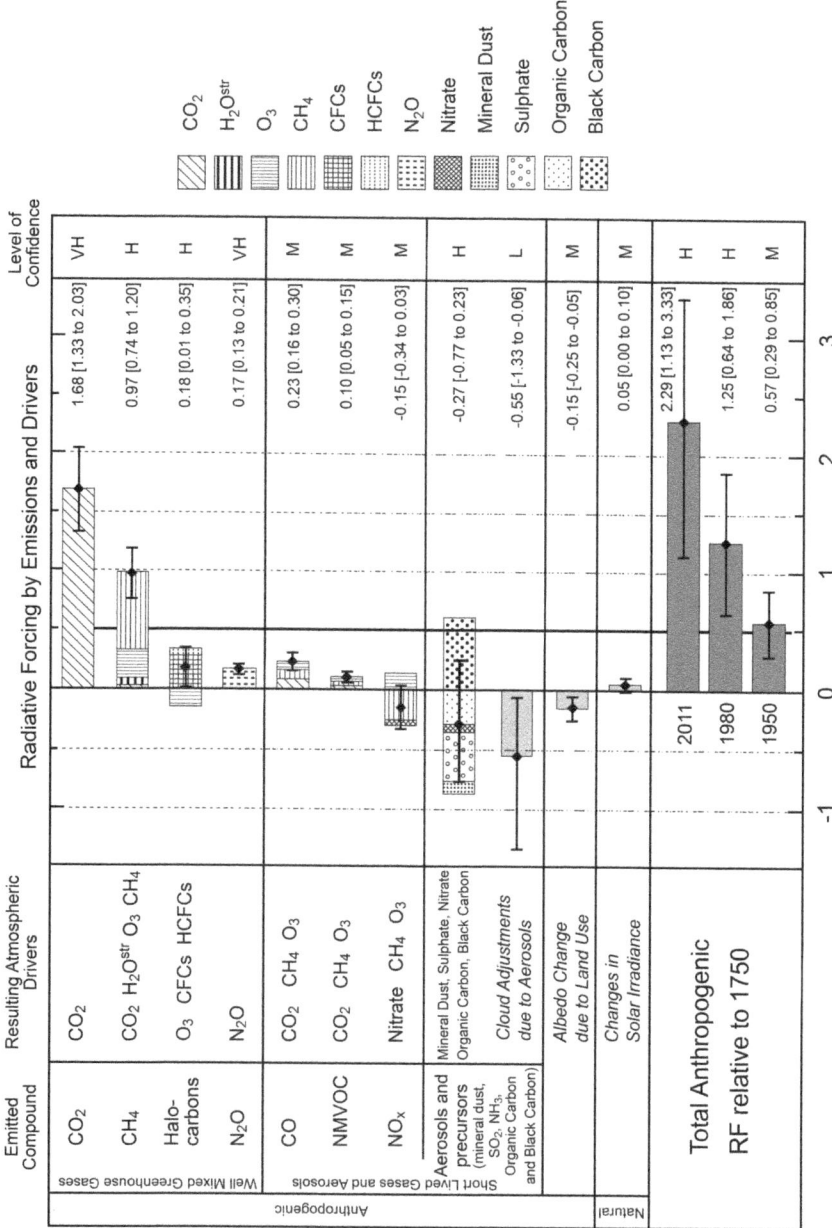

Figure 4 Radiative forcing by emissions and other drivers (IPCC 2013).[9]

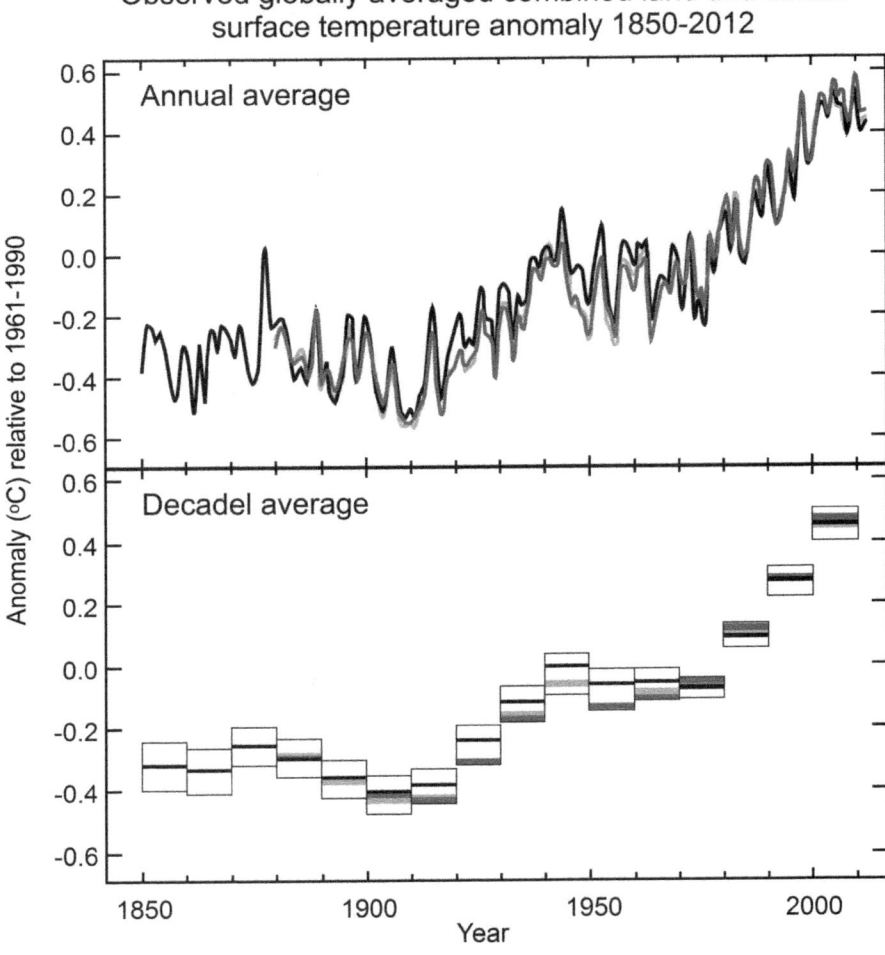

Figure 5   Observed globally averaged combined land and ocean surface temperature anomaly 1850–2012 (Data from the IPCC 2013).[9]

Over the last two decades, the Greenland and Antarctic ice sheets have been losing mass, glaciers have continued to shrink almost worldwide, and Arctic sea ice and Northern Hemisphere spring snow cover have continued to decrease in extent.[9] (p.7)

This has been most noticeable in the Arctic where annual sea ice extent has shrunk considerably, especially in summer. The picture is less clear in Antarctica where some areas are seeing growth of ice extent where other areas are seeing a retreat. Permafrost temperatures have increased and in parts of the Russian European North reductions in permafrost thickness and extent have been observed since the 1970s.

The rate of sea level rise since the mid-nineteenth century has been larger than the mean rate during the previous two millennia. Over the period 1901–2010 global sea level rose by 0.19 (0.17–0.21) m.[9] (p.9)

Ocean thermal expansion and glacier mass loss together explain about 75% of the observed global sea level rise since 1970. In the last interglacial, 129 000 to 116 000 BP (before present years), sea level was between 5 and 10 m above present levels.

The atmospheric concentrations of carbon dioxide (CO_2), methane, and nitrous oxide have increased to levels unprecedented in at least the last 800 000 years. CO_2 concentrations have increased by 40% since pre-industrial times, primarily from fossil fuel emissions and secondly from net land use change emissions. The ocean has absorbed about 30% of the emitted anthropogenic carbon dioxide, causing ocean acidification.[9] (p.9)

Figure 6 shows that the total emissions of carbon dioxide and the global mean surface temperature response are approximately linearly related. In order to limit global warming to less than 2 °C it is shown that anthropogenic emissions from all sources must not exceed approximately 1 trillion

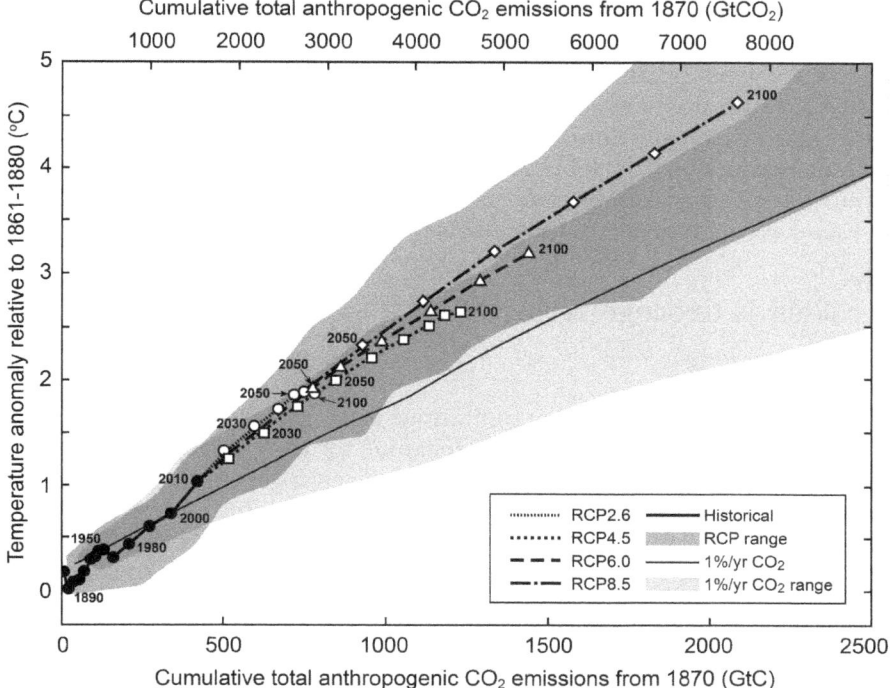

Figure 6   Cumulative total anthropogenic CO_2 emissions from 1870 (http://www.climatechange2013.org/images/figures/WGI_AR5_FigSPM-10.jpg).

tonnes of carbon. Emissions are continuing to rise despite the recent global recession and we are already halfway towards this target. The carbon dioxide that is already in the climate system will continue to raise temperatures even if the world runs on carbon-free energy.

> *A large fraction of anthropogenic climate change resulting from CO_2 emissions is irreversible on a multi-century to millennial time scale, except in the case of a large net removal of CO_2 from the atmosphere for a sustained period. Surface temperatures will remain approximately constant at elevated levels for many centuries after a complete cessation of net anthropogenic CO_2 emissions. Due to the long time scales of heat transfer from the ocean surface to depth, ocean warming will continue for centuries. Depending on the scenario, about 15 to 40% of emitted CO_2 will remain in the atmosphere longer than 1000 years.[9] (p.26)*
>
> *There are many approaches and pathways to a sustainable and resilient future. However, limits to resilience are faced when thresholds or tipping points associated with social and/or natural systems are exceeded, posing severe challenges for adaptation.[10] (p.20)*

Clearly we need new solutions to global warming since carbon dioxide induced warming will continue for centuries even if there is a complete cessation of all emissions. This increase in energy to the climate system will increase the severity of extreme events leading to unprecedented heat waves, floods, droughts, storms and subsequent landslides, tsunamis and coastal erosion. These have knock on impacts on humans *via* agriculture, food security, forestry, health, infrastructure, tourism, transport and water supply that may cause a host of global problems including increased poverty, reduced security, increased migration and possibly a breakdown of society as we know it. Could geoengineering of the climate system provide this new solution?

4   What is Geoengineering?

4.1   Introduction

Many different geoengineering definitions exist, and it is interesting to observe the solidification of the word's meaning through time. See Box 1 which provides the *Oxford English Dictionary* reference for geoengineering.[11] It can be seen that since approximately the 1980s, a reasonably consistent idea of geoengineering has settled in the English language. The influential 2009 Royal Society report on geoengineering succinctly defines geoengineering as "the deliberate large-scale intervention in the Earth's climate system, in order to moderate global warming" and this definition is used henceforth.[12] There is currently a debate about whether the term 'geoengineering' should be replaced with 'climate engineering' since it offers clearer labelling. However, there is little evidence at present that geoengineering is being usurped by climate engineering, within the British media at least.

Box 1   OED definition of geoengineering.[11]

The Oxford English Dictionary documents changes in the meaning and use of words throughout history.

Geoengineering, noun.

Definition – The application of large-scale engineering methods to modify rock formations or other features of the natural environment; (in later use esp.) the modification of the global environment or the climate in order to counter or ameliorate climate change.

1962 *New Mexican (Santa Fe, New Mexico)* 16 Feb. 9/3. The 30 graduate geologists currently employed by the department, either as materials testers or as central research laboratory workers handling geoengineering assignments.

1969 *Sci. News* **95** 159/1. Teller... urged that Australia act as a proving ground for nuclear geoengineering.

1976 C. Marchetti, *Geoengineering & CO₂ Probl.* p. iii. Geoengineering... is a kind of 'system synthesis' where solutions to global problems are attempted from a global view.

1983 T. Hoyle, *Last Gasp* iv. 51. The Russians are keen to find out everything they can about what affects the climate because of their grandiose geoengineering schemes.

1994 *Guardian* 17 Mar. 94. The market can supply appropriate geoengineering—for example, companies launching mirrors into space to deflect sunlight.

2007 *Nature* 10 May 115/2. Geoengineering... explores in what circumstances aspects of the climate system might be deliberately modified to limit the worst eventualities of climate change.

Geoengineering is distinct from climate change mitigation and adaptation. Mitigation involves strategies to reduce anthropogenic emissions of greenhouse gases, for example through the transition to a low carbon economy. Adaptation aims to increase resilience towards the effects of climate change, for example by improving flood defences and providing protection against climate dependent disease. Geoengineering is not a new idea and it has its origins in weather modification studies.[13] A good history of geoengineering, as opposed to weather modification, is provided by the review of Keith.[14]

Geoengineering schemes can be divided into two main categories, namely: Carbon Dioxide Removal (CDR) and Solar Radiation Management (SRM), and the subsequent chapters in this book provide specific details about various CDR and SRM schemes: Chapters 2 and 3 – carbon sequestration (CDR); Chapter 4 – artificial trees (CDR); Chapter 5 – increased surface albedo; Chapter 6 – brighter clouds (SRM); Chapter 7 – stratospheric aerosol (SRM); and Chapter 8 – space based solutions (SRM).

Geoengineering schemes typically aim to reduce climate change impacts through use of analogues of processes already present in the Earth system. For example stratospheric particle injection (see Chapter 7) mimics the effect of large volcanic eruptions; these eruptions can inject megatonnes of particle material into the stratosphere leading to an increased global albedo and planetary cooling. The last eruption to have a major effect on climate was Mt Pinatubo in the Philippines in 1991 which reduced global surface temperatures by ~ 0.5 °C for a 1–2 years.[15] The geoengineering analogue to volcanic activity is anthropogenic particle injection into the stratosphere *via* a non-volcanic route such as an aeroplane or pipe delivery system.[16]

CDR technologies aim to reduce climate change by removing greenhouse gases from the atmosphere. The greenhouse gas of choice is typically CO_2 because its high concentration makes extraction easier. Different CDR techniques use biological, chemical or physical approaches to remove CO_2 from the atmosphere. CDR is likely to be slow, compared to SRM, because of the huge amounts of CO_2 that need to be removed. The *IPCC 5th Assessment Report* (AR5) evaluates that CDR techniques would need to be deployed at large scale for over a century to be able to significantly decrease CO_2 concentrations.[9]

SRM schemes decrease the effects of climate change by reducing the amount of energy within the Earth system by reflecting a proportion of solar radiation back to space. It is expected that SRM techniques, if technologically feasible, would be able to quickly decrease average global temperatures over the timescale of a decade or so. Unlike CDR, SRM does not remove greenhouse gases from the atmosphere and, as such, can only be viewed as a temporary solution to climate change, albeit one which might be used to buy time whilst civilization transfers to a low carbon economy, or successfully integrates CDR schemes to counterbalance fossil fuel burning. Another obvious downside of SRM is that it only counteracts the radiative effects of greenhouse gas emissions; it does not counteract the non-radiative effects such as the partitioning of excess atmospheric CO_2 into the oceans. Hence ocean acidification would still occur under SRM scenarios and itself represents a serious environmental threat.

Another potential issue with SRM geoengineering is the termination problem. Since SRM could rapidly decrease the Earth's temperature, if it is turned off the restoration of climate to its non-geoengineered state would also be fast. Rapid rises in temperature are much more stressful for ecosystems and human infrastructure to cope with than slow changes. If SRM techniques were found to be causing unplanned and deleterious effects the termination problem would then cause a dilemma: rapidly switch off SRM and risk stressing various systems, or gently ramp down SRM thus prolonging the unwanted side effects. CDR techniques are not as susceptible to termination risks because of their slower timescale of action.

The relative ranking of different geoengineering proposals is difficult because of the multiple assessment criteria. It is not clear at present what the best metrics are with which to assess geoengineering proposals. The influential Royal Society report uses four main technical criteria through which to rank the different technologies: effectiveness, timeliness, safety and affordability.[12] The report highlights that no single technology, identified to date, scores highly in all four sections. Moreover there are also non-technological criteria which should be used to assess the different technologies; these are much more difficult to rank quantitatively and include: public attitudes, social acceptability, political feasibility and legality. The complexity further increases with many of the ranking criteria liable to change over time.[12]

It should be noted that, if used, geoengineering does not have to be employed to completely remedy all of anthropogenic climate change. It might be more usefully utilized as part of a toolbox of technologies and policies with which to stabilize climate change.[17]

4.2   Are there Parallels to Climate Change and Geoengineering?

Myriad environmental risks and problems have been created since civilization entered into the Anthropocene – the human-dominated geological epoch that overtook the Holocene.[18] In fact it is likely that geologists of the future will use these anthropogenic changes to exemplify the Anthropocene. In addition to anthropogenic climate change, these environmental problems and risks include but are not limited to: stratospheric ozone depletion; biodiversity loss; overfishing; changes in the phosphorus and nitrogen cycles; and chemical pollution including nuclear waste and persistent organic pollutants.[19] Many of these environmental pollutant problems can be understood in the context of a tragedy of the commons analysis,[5] in which a common resource (*e.g.* the atmosphere and oceans) is overused to individual advantage but to the group's disadvantage.

Global problems of pollution typically arise because of the persistence of pollutants and the limited ability of natural systems to absorb anthropogenic stressors. If the lifetime of a pollutant is sufficiently long then the pollutant can be transported globally. For atmospheric pollutants, a lifetime greater than a couple of years will lead to global ubiquity and persistence. The atmospheric lifetime of CO_2 is difficult to define exactly due to multiple location dependent loss processes, such as uptake by the oceans and biosphere, but it is approximately on the timescale of fifty to a hundred years. This makes CO_2 a cross national boundary pollutant, and defines a molecule of CO_2 released in one specific country as dangerous as a molecule released in any other country. The cross boundary nature of CO_2 can be viewed as a benefit for CDR approaches to geoengineering. If the significant removal of the long lived CO_2 gas can be achieved, then the technology can be deployed anywhere worldwide to reduce the global burden of atmospheric CO_2.

It is perhaps useful to investigate whether there are contemporary or historic parallels to climate change and geoengineering. The identification of the Antarctic ozone hole and the subsequent phase-out of ozone destroying chemicals is often cited as the paradigm of the success of environmental science. Furthermore it is also often used as evidence that anthropogenic climate change can be fully understood and rectified.

The Antarctic ozone hole was first identified in the 1985 and it was soon understood that the release of halogen containing species, in the form of refrigerants and propellants, was leading to stratospheric ozone depletion especially in the polar regions.[20] Once the cause and effect of the ozone hole had been largely understood, then international treaties, starting with the *Montreal Protocol on Substances that Deplete the Ozone Layer* (1987), rapidly started to reduce the concentration of ozone destroying substances in the stratosphere. A mitigation approach to the ozone hole problem was successfully instigated and carried out. Geoengineering or adaptation type approaches were not used or required. Due to these efforts, the ozone hole is now expected to completely recover by approximately the year 2065.

So now that the scientific understanding of climate change is maturing and attribution of anthropogenic emissions to the warming of the climate system is explicit,[9] could a similar strategy to that used with the ozone hole problem be followed? The major difference between the ozone hole and climate change is the transferability of the responsible pollutant. The global economy was not reliant on ozone depleting refrigerants and propellants and hence the mitigation approach was relatively straight forward. For the ozone hole problem, it was somewhat simple to find replacement refrigerants and propellants which possessed much lower ozone depleting potentials. By contrast CO_2, the main greenhouse gas, has no substitute, and if we continue to burn hydrocarbons then CO_2 will result. At present the global economy and its infrastructure are reliant on burning hydrocarbons.

Whilst climate change has clear parallels with other environmental problems, including the ozone hole, it is the largest environmental problem that the human race has so far encountered. It is global in scope and the chief greenhouse gases responsible cannot be substituted for less harmful chemical species. As such it is probably the most difficult, as well as the largest, environmental problem encountered.

4.3 Scientific Respectability of Geoengineering

The IPCC has for the first time described geoengineering in its most recent fifth assessment report.[9] Whilst the report uses rather bland language, the inclusion of geoengineering within this document constitutes a coming of age moment for the science of geoengineering which in many eyes has previously looked like nothing more than crank science.

A milestone for the acceptance of geoengineering into the global scientific discourse on climate change occurred upon publication of an editorial by Paul Crutzen, who was jointly awarded the Nobel Prize in Chemistry in 1995 for his work in atmospheric science.[21] Within this editorial the SRM technique of stratospheric injection of aerosols was evaluated and it stated "… although by far not the best solution, the usefulness of artificially enhancing earth's albedo and thereby cooling climate by adding sunlight reflecting aerosol in the stratosphere might be explored and debated…". This cautious justification of geoengineering research by one of the luminaries of environmental science helped to give legitimacy to the nascent field of geoengineering.

More recently, another prominent and highly respected scientist providing further cautious justification of research has been Lord Rees of Ludlow who has filled several of the top scientific positions within UK science including: presidency of the Royal Society, Astronomer Royal, and Master of Trinity College, Cambridge. "Most nations now recognize the need to shift to a low-carbon economy, and nothing should divert us from the main priority of reducing greenhouse gas emissions. But if such reductions achieve too little too late, there will surely be pressure to consider a plan 'B' – to seek ways to counteract the climatic effects of greenhouse gas emissions by 'geoengineering'."[12]

The greater visibility of geoengineering science, in part helped by the respectability given by such examples, combined with the bleak outlook on climate change has led to an explosion in research output on geoengineering (see Figure 7). This output has been spread over a wide range of disciplines including the physical and social sciences, and engineering.

4.4 The Arguments for and against Geoengineering Research

Geoengineering *via* CDR techniques is far less controversial than SRM techniques and as such there is little opposition to research and development of CDR techniques. Conversely there has been loud and fierce

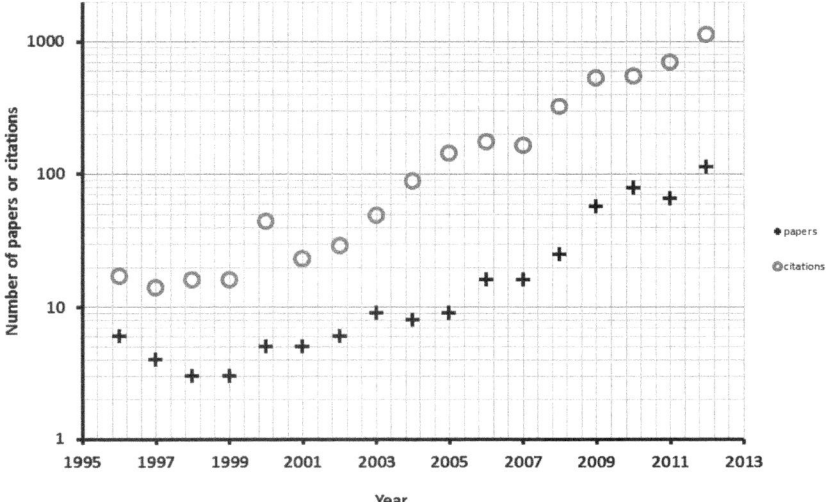

Figure 7   Geoengineering research output.
(Data compiled from the search returns to the subject query 'geoengineering' within Scopus, an abstract and citation database of peer-reviewed literature; www.elsevier.com/online-tools/scopus).

opposition to the research and development of SRM techniques. Most SRM research has been justified by framing it in the context of future proofing society against unbearable climate change. The argument is as follows: mitigation by reducing global greenhouse emissions is the best way to solve climate change because it keeps the planet closer to its natural state. However, in the last decades there has been little progress on reducing emissions and in fact our emissions are increasing. For example our CO_2 emissions are rising by the equivalent of approximately 2 ppmv per year at present (see Figures 2 and 3). Therefore there are justified concerns and wide spread pessimism about our ability to curb emissions in the short term (*ca.* 100 years). This generates real worries that civilization might not be able to mitigate CO_2 in time, hence geoengineering might be required. This argument frames geoengineering as the lesser of two evils when compared to dangerous climate change, and it follows that if there is the possibility that we will need geoengineering in the future, then we should prepare for that possibility now.

There are clearly different levels of environmental risk associated with the three major strands of geoscientific research: modelling, laboratory and field studies. Desk studies that use computer models to predict the outcomes of geoengineering offer zero risk to the environment. Likewise laboratory studies offer virtually no risk to the environment since they do not interact with world external to the laboratory. Field trials are more problematic and could interact with biogeochemical cycles if their scale was large enough. Common sense should dictate that if and when field trials are attempted, they would start at sufficiently small scales so their environmental effect would be localised in time and space, and negligible over wider temporal and spatial regions. If initial small scale trials were successful then increasingly larger and longer trials would be implemented with the increased risk of environmental harm.

The future proofing rationale for geoengineering research has many opponents both individual and institutional. The arguments raised against geoengineering research are numerous, and will not be fully explored here, but they can broadly be classified as technical, moral or political in standpoint.

From the technical standpoint, a major worry about geoengineering research is: how can you test and validate the various geoengineering schemes? The Earth is a highly complex non-linear system with numerous feedbacks which are not fully understood. Whilst various geoengineering schemes could be simulated within computer models and laboratory tests, at some stage field trials would be required. Validation would require field trials of sufficient size to generate a response in the Earth's climate so a clear cause and effect could be observed. This would necessitate large field trials.

Geoengineering would require huge infrastructure and with that infrastructure there would be the possibility of human error. Recently there have been two large examples of large negative impacts of human error on the environment: the Deepwater Horizon oil spill (2010) and the Fukushima

Daiichi nuclear disaster (2011). All technical validations come with a level of uncertainty. What level of uncertainty would be acceptable for geoengineering? There is also the worry of unanticipated consequences of geoengineering – the so called 'unknown' unknowns. What is the likelihood of these being foreseen prior to a major scale geoengineering attempt? The precautionary principle might seem to argue strongly against geoengineering.

From the political standpoint, there is a suspicion that geoengineering will encourage political inertia with respect to CO_2 mitigation. A situation of moral hazard could be generated in which geoengineering provides a political get out clause for the more costly option of mitigation. Hence if geoengineering technologies are developed and are ready to be deployed, then there is less chance that mitigation will occur in a significant and timely fashion. Whilst geoengineering research is in its youth and so far its research costs have been minor, is geoengineering the correct big project for governments to put their money into? Would geoengineering research remove money from the mitigation research effort? The long lifetime of greenhouse gases and the climate change problem in general, makes legislation difficult due to the time inconsistency between the timescale of government (years to decades) and the timescale of climate change (decades to centuries).

The governance of geoengineering is another problem. Who would own the technology? And who would fund the process? It seems likely that if geoengineering is to happen then the UN would have to be the ultimate responsible authority. At present, there are no specific UN based regulations against geoengineering. However, several existing frameworks are thought to have relevance to some types of geoengineering, for example: *The Convention on the Prevention of Marine Pollution by Dumping of Wastes and Other Matter* (1972) might be applied to iron fertilization of the oceans, and *The Environmental Modification Convention* (1978) could potentially be applied more widely to geoengineering schemes in general. In common with the effects of climate change. the effects of geoengineering are likely to be unevenly distributed globally. There will be winners and losers, so how is equity ensured amongst the nations? Would compensation be required for the losers? Geoengineering would need a final goal – a climatic endpoint. Who sets this goal? Different countries and regions will clearly have different views on what constitutes a desirable outcome. Moreover is it possible that some countries would regard global warming to be a positive outcome, for example, in polar regions? Geoengineering is likely to be cheap compared to the transfer of the world to a low carbon economy, at least in the near future. Therefore it will be financially possible for large countries and companies. How do we avoid unilateralism?

Geoengineering offers moral questions as well. Such as: is geoengineering hubristic (see Chapter 9)? Does this technology represent a step too far in our control of the Earth system? By even talking about geoengineering, do we normalize the concept or provide it with premature credibility? Does the

mere presence of geoengineering research represent a slippery slope towards
the ultimate use of geoengineering? This is the atomic bomb argument –
that once a technology has been developed it is unlikely that it will not be
utilized.

5   Summary and Conclusions

The atmosphere is the most valuable resource on the planet and as such
every effort needs to be made to protect it. Unfortunately the rise in greenhouse gases since the industrial revolution, and the intimately linked
change in climate, is proving to be the most difficult environmental problem. Even though the strongest scientific evidence tells us that the anthropogenic releases of greenhouse gases are responsible for climate change,
there has been little success in emissions reduction. The reasons behind this
failure are complex but the outcome is not; the regions of the Earth inhabited by humans are on average getting hotter.

Since mitigation efforts against climate change are failing, the arguments
for the possibility of geoengineering become louder. Geoengineering is a
contentious issue which evokes strong reactions within all levels of society.
SRM technologies are more controversial than CDR technologies since they
do not solve the root cause of the problem, however, they potentially offer a
more rapidly deployed solution. At present no geoengineering technology is
fit for purpose or ready for deployment. However, geoengineering research is
rapidly increasing with hundreds, if not thousands, of scientists and engineers working on the topic worldwide. As such, geoengineering research
has now likely passed through its infancy, and conclusions are being
reached about the efficacy, benefits and disadvantages of the different proposals. It seems increasingly likely that geoengineering technologies could
be developed that will reduce the effects of climate change. These benefits
need to be carefully weighed against the detriments. A true assessment of
geoengineering cannot be achieved until we better understand the environmental, technological, economic and governance issues, associated
through its use. Thus this book sets out our present knowledge of this
subject, as well as the areas where understanding is currently lacking.

References

1. NOAA's Mauna Loas Observatory sees 400 ppm carbon dioxide
   levels, *Planet Save*; http://planetsave.com/2013/05/14/noaas-mauna-loa-
   observatory-sees-400ppm-carbon-dioxide-levels/ (last accessed 30th
   October, 2013).
2. B. Holmes, *New Scientist*, 2013, **2936**, 38–41.
3. J. Lovelock, *Gaia, A New Look at Life on Earth*, Oxford University Press,
   Oxford, 1979.
4. J. Lovelock, *The Revenge of Gaia*, Penguin, London, 2006.

5. G. Hardin, *Science*, 1968, **162**(3859), 1243–1248.
6. J. Rogelj, D. L. McCollum and K. Riahi, *Nature Climate Change*, 2013, **3**, 545–551.
7. J. E. Thornes, in *Ecosystem Services*, ed. R. E. Hester and R. M. Harrison, Royal Society of Chemistry, Cambridge, 2010, 70–104.
8. R. A. Silva, J. J. West, Y. Zhang, S. C. Anenberg, J.-F. Lamarque, D. T. Shindell, W. J. Collins, S. Dalsoren, G. Faluvegi, G. Folberth, L. W. Horowitz, T. Nagashima, V. Naik, S. Rumbold, R. Skeie, K. Sudo, T. Takemura, D. Bergmann, P. Cameron-Smith, I. Cionni, R. M. Doherty, V. Eyring, B. Josse, I. A. MacKenzie, D. Plummer, M. Righi, D. S. Stevenson, S. Strode, S. Szopa and G. Zeng, *Environ. Res. Lett.*, 2013, **8**, 034005.
9. IPCC, *WG1 Summary for Policymakers*, 27 September 2013, http://www.ipcc.ch/report/ar5/wg1/docs/WGIAR5_SPM_brochure_en.pdf.
10. IPCC, Summary for Policymakers, in *Managing the Risks of Extreme Events and Disasters to Advance Climate Change Adaptation*, A Special Report of Working Groups I and II of the IPCC, Cambridge University Press, 2012, pp. 3–21.
11. *Oxford English Dictionary*; http://www.oed.com/ (last accessed 6th January, 2013).
12. J. Shepherd, *Geoengineering the Climate: Science, Governance and Uncertainty*, The Royal Society, 2009; http://royalsociety.org/policy/publications/2009/geoengineering-climate/ (last accessed 6th January, 2013).
13. J. R. Fleming, *Fixing the Sky,* Columbia University Press, New York, 2010.
14. D. W. Keith, *Annu. Rev. Energy Environ.*, 2000, **25**, 245–284.
15. A. Lambert, R. G. Grainger, J. J. Remedios, C. D. Rodgers, M. Corney and F. W. Taylor, *Geophys. Res. Lett.*, 1993, **20**(12), 1287–1290.
16. F. D. Pope, P. Braesicke, R. G. Grainger, M. Kalberer, I. M. Watson, P. J. Davidson and R. A. Cox, *Nature Climate Change*, 2012, **2**, 713–719.
17. S. Pacalaw and R. Socolow, *Science*, 2004, **305**, 5686.
18. P. J. Crutzen, The "Anthropocene", in *Earth System Science in the Anthropocene*, ed. E. Ehlers and T. Krafft, Springer, Berlin, 2006.
19. J. Rockström, W. Steffen, K. Noone, Å. Persson, F. S. Chapin, E. F. Lambin, T. M. Lenton, M. Scheffer, C Folke, H. J. Schellnhuber, B. Nykvist, C. A. de Wit, T. Hughes, S. van der Leeuw, H. Rodhe, S. Sörlin, P. K. Snyder, R. Costanza1, U. Svedin, M. Falkenmark, L. Karlberg, R. W. Corell, V. J. Fabry, J. Hansen, B. Walker, D. Liverman, K. Richardson, P. Crutzen and J. A. Foley, *Nature*, 2009, **461**, 472.
20. WMO (World Meteorological Organization), Scientific Assessment of Ozone Depletion: 2010, *Global Ozone Research and Monitoring Project – Report No. 52*, Geneva, Switzerland, 2011, p. 516.
21. P. J. Crutzen, *Climatic Change*, 2006, 77(3–4), 211–220.

Storing Carbon for Geologically Long Timescales to Engineer Climate

R. STUART HASZELDINE* AND VIVIAN SCOTT

ABSTRACT

To re-establish global climate balance, it is necessary to remove large amounts of fossil carbon emitted by humans, which is currently located in the atmosphere and the upper ocean. Although great attention is given to technologies of capture, the ability to store immense tonnages of carbon stock for geologically long time periods, isolated from atmosphere and ocean interaction, is equally important. In this chapter, the multiple storage locations for carbon stocking on and below land, also within and below the ocean, are evaluated. The evaluation shows that carbon dioxide reduction (CDR) is useful for mitigation, but cannot balance the rate of new emissions from fossil fuel exploitation. Many CDR methods have large uncertainty in their quantity, life-cycle, global impact and engineered feasibility. Competition for biomass and land usage is inevitable. Pathways and reservoirs of carbon in the ocean are complex and interlocked. Engineered storage of carbon will also be expensive, resource intensive and cannot substitute for a greatly reduced usage of fossil carbon. Human industrial and economic activity must "move beyond hydrocarbons" to be sustainable beyond 2050.

---

*Corresponding author

Issues in Environmental Science and Technology, 38
Geoengineering of the Climate System
Edited by R.E. Hester and R.M. Harrison
© The Royal Society of Chemistry 2014
Published by the Royal Society of Chemistry, www.rsc.org

1 Why is Carbon Storage Necessary?

Industrial and advanced societies gain a large part of their energy usage from combustion or dissociation of fossil carbon: coal, oil, and methane. Each one of these has augmented the previous energy vector, but not replaced it. Thus, more wood is combusted today than ever before, as well as more coal and more oil. Although there is certainly a trend towards more hydrogen rich fuels, now expressed as the rise of methane usage, the overall release of fossil carbon from the geosphere into the atmosphere and ocean has proved relentless. The historical rise in carbon emissions continues to accelerate, notably since the industrial revolution, founded in the UK, made condensers so that coal-fuelled steam engines became more efficient and commercially valuable. The rest is, literally, history.

The immediate consequences of utilisation of fossil carbon were, and still are, vastly improved wealth, leading to improved human health, fertility, and long life. These are all highly desirable, and the activity of consuming fossil energy directly, or by proxy, permeates our entire culture. However, it is now clear that the carbon emissions from these activities have overwhelmed the natural processes for circulation of carbon through the Earth's atmosphere, biosphere, and ocean, such that humans now dominate the inelastic parameters of the global carbon cycle. That enables naming of our present geological epoch the "Anthropocene",[1] where humans dominate over many Earth processes. Although Arrhenius realised theoretically that carbon dioxide is able to retain solar heat reflected from Earth in the atmosphere,[2] the consequences have only gradually become apparent with the empirical measurements of Keeling,[3] leading to seminal predictions of climate warming by Hansen in 1981.[4] Thirty years of focused and intense scientific investigation has led to a much improved understanding of carbon cycling on Earth, and the contribution that CO_2 makes to the greenhouse gas in the Earth's atmosphere.

There is intense debate concerning the rates of CO_2 increase or the rates of carbon production and the rates of temperature change. However, in a geological context, all these timescales are instantaneous. During the past 250 years, humans have released more than one quarter of the carbon emitted by the Earth during some of its more intense volcanic episodes during the past 600 million years. It is now clear from climate modelling that the rate is not so important, but the total quantity of carbon released is a fundamental controlling factor of global change.[5] The conclusions of climate modelling are supported by the geological record. Although the resolution in calendar years, or hundreds of thousands of years, is poor in comparison with the most recent 100 000 years, it is clear that the Earth has emitted large quantities of carbon dioxide at perhaps five different times during the past 600 million years. At each of those times there has been a rapid period of global change, often associated with temperature warming, which has resulted in geologically instantaneous extinctions of many species. Consequently, the emission of fossil fuels can be viewed as a carbon stock problem, not a rate problem. The present emissions can be viewed as an experiment in undertaking the rapid release of CO_2 during geological time, and thereby experimenting with a sixth extinction.

A logical deduction from these observations is that humans have a vested interest in managing the carbon stock on the Earth's surface. That includes reducing the total quantity of carbon stocked rapidly into the atmosphere and ocean during the past 250 years. If carbon can be captured using diverse natural or engineered processes, where then can this carbon be stored? It is clear from consideration of the Earth as a system, that this carbon has to be stored not just for one year or 100 years, but for thousands or tens of thousands of years whilst the Earth's natural self regulation returns to a level with which we are familiar during the past 15 000 years, after the last de-glaciation. If humans are not able to find methods to reduce the carbon stock, and store carbon for these extended time periods, then the Earth will continue to undergo a global change, similar to many in the geological past, where present climate belts move pole-wards, and the familiar pattern of ocean currents, seasonality, rainfall, temperature, and weather becomes unpredictable. In climate modelling, the threshold at which these adverse effects become unpredictable has conventionally been taken at 2 °C. That is not necessarily a hard boundary, but a clear indication of a threshold, after which reversibility becomes much more conjectural. Taking the analysis of Meinshausen,[5] that 1 000 000 000 000 tonnes of carbon is the regulated total amount, Allen has calculated that this threshold will be reached in 2044.[6] Humans have a large stake in correctly managing the carbon stock before that date is reached.

In the remainder of this article, we assess the different styles by which carbon may be stored, and analyse the known information, the natural processes, and the engineering interventions which could be undertaken to enhance the global rate and tonnage of carbon stocking.

2   The Approach and Controlling Factors

Multiple methods to reduce carbon dioxide, and to store carbon dioxide or carbon, have been proposed during the past 30 years, and it is likely that additional methods will be proposed in the future. Here we categorise the methods which seem, in our analysis, to have the largest global potential. We subject each of these methods to a similar suite of analysis. Firstly we describe the essential features of the method, in terms of its process, geography, and chemistry; we assess its potential impact in terms of global tonnage of CO_2 per year; then we attempt to estimate the cost, engineering feasibility, security of carbon retention through centuries or millennia, the effort needed to maintain that carbon stock through our millennial time-scale; and finally we speculate on the potential adverse effects which may result from the method's adoption.

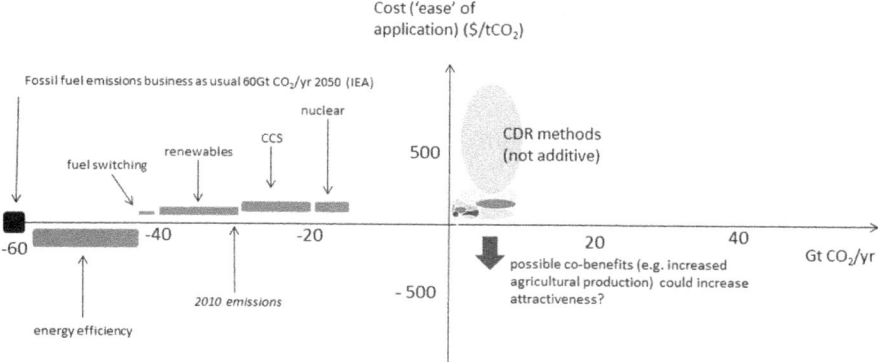

Figure 1   Diagram showing the size of the challenge for engineered carbon production and storage.

To gain some insight into the significance, or impact, of the method, these
different opportunities are portrayed in Figure 1 both graphically and nu-
merically, in relation to the present day production of fossil carbon recorded
by organisations such as BP,[7] or the International Energy Agency.[8]

Figure 1 illustrates the degree of the challenge facing engineered carbon
production and storage. The X axis shows different quantities of CO_2
emission to the left, with diverse possibilities of CO_2 storage to the right. The
left black square is the total amount of fossil carbon dioxide emitted each
year by 2050 if a 'business as usual' trajectory is followed, according to the
International Energy Agency. The series of ellipses to the right of the y-axis
show different possibilities of climate engineering to store CO_2. Horizontal
dimension estimates the range of CO_2 storage per year, the vertical dimension estimates the range of cost per tonne of CO_2 which that action may
incur. It is clear that no single action of carbon storage is anywhere close to
being sufficient to balance the fossil CO_2 emitted annually. Additional
measures are needed.

The blue horizontal bars estimate the tonnage of CO_2 reduction which
could be achieved by different measures which do not involve climate engineering. Bars below the horizontal $(Y = 0)$ are actions which save money,
and are in principle self funding. Bars above the horizontal are actions
which cost money, expressed as $ per tonne of carbon dioxide on the Y axis.
Even if all five conventional actions are followed, there is still a net CO_2
emission of around 15 Gt CO_2 yr^{-1}. From this, it remains unclear whether
even if all carbon storage engineering actions are taken, they would be
capable of balancing the residual CO_2 emissions, especially in the long term.
The fundamental conclusion from this diagram is that consumption of fossil
carbon, and the rate of emission, has to be curtailed in ways which are not
yet calculated. Climate engineering by storage of fossil carbon emitted
cannot balance the projected rates of new emission, and therefore cannot,
on its own, manage the stabilizing of atmospheric CO_2.

3   Methods of Reduced Emission Rates

There are methods which, in principle, are well understood to enable reduction of fossil carbon emissions. Many of these reside in the domain of
energy studies. For example switching of fuel vector from coal, to oil, and
then to methane can produce similar amounts of energy with systematically
decreased carbon emission. This relies on the oxidation of hydrogen to form
H_2O as the main exothermic route. It is well understood that many nations
globally depend on combustion of large amounts of coal to produce electricity,
and also heat. Examples include China, USA, Germany, and the UK. Coal is
both cheap and available. Environmental policies to reduce its use have been
ineffective since the 1980s, in spite of the large costs of its extraction, in terms
of human life and in direct environmental degradation. More successful, has
been the economic case for fuel switching. Since 1990 in the USA, unanticipated discoveries of large resources of shale gas have produced a glut of low-
cost fossil hydrocarbon into the USA domestic market. This has out competed
coal on its low price, and has led to a decrease in carbon emissions whilst
maintaining a similar pattern of energy usage. However, it is clear from basic
calculations of energy modelling that switching from coal or oil to gas only
buys two extra decades of time before the total global carbon budget is exceeded. More aggressive and even lower carbon methods are required.

A favourite method for many political and industrial leaders is that offered
by carbon capture and storage (CCS). This allows continued combustion of
coal, oil, or gas at electricity generation plant, or industrial process sites.
Emitted CO_2 is avoided by chemical transformation of fuel pre-combustion,
or the CO_2 is chemically adsorbed from flue gas after combustion. The pure
CO_2 is liquefied, transported by pipeline, and injected 1 to 4 km below
ground where it can remain for millennia. This can reduce carbon emissions
at industrial sites by 90%. However, the initial projects have proven to be too
expensive for national governments to implement, and insufficiently cour-
ageous environmental legislation has been produced to enforce market
companies to implement CCS. If this group of technologies does eventually
become deployed, then that too can buy an extra 20 to 40 years of time until
the global carbon budget is exceeded.

It is also well understood that energy usage by industrial societies and
individuals is inefficient. If the price of energy were perhaps 10 times its
present level then it would be commercially worthwhile to develop district
heating schemes, insulate houses, share bus transport rather than use individual cars, or use centralised electricity power plants which have a thermal efficiency greater than 35%. Again, the principles are well known, but
the delivery has been perennially slow through many decades. Assertively
explained lifestyle changes, predicated around energy efficiency, could reduce consumption rates by 30 to 70%.[7,9] That could buy perhaps another 60
years of time before the carbon budget is exceeded.

In the event that none of these above methods are implemented sufficiently, at scale and globally, then to stabilize climate, humans will be

required to undertake deliberate climate engineering to undertake solar radiation management (SRM), or to undertake deliberate carbon extraction and storage from the biosphere atmosphere and ocean. These technologies are analysed in the following sections.

4 Principles of Carbon Dioxide Removal (Negative Emissions Technologies)

Reduced rates of CO_2 emission cannot, on their own, keep the industrialised world within its global carbon budget. Either it will be necessary to undertake several, or all, of the above emissions reductions, or additional technology interventions will be required. Carbon dioxide removal (CDR), or negative emissions technologies,[10] undertake a deliberate removal of CO_2 from the atmosphere and ocean, sending it to storage and isolation for thousands of years. In the compilation below we analyse the arithmetic claims for CDR technologies. The real effects of CDR will be more complex. For example if CO_2 is reduced in the atmosphere, then that will enable additional CO_2 release from the upper ocean.[11] As a second example, increasing forest cover in the tropics may draw down atmospheric CO_2, but this will consequently reduce the rate of vegetation growth. Precise predictions will require much more detailed modelling and life-cycle analysis of interacting natural processes.

5 Life-cycle Assessments

The size of the problem involves very large tonnages of CO_2. Consequently, the development of many of these terrestrial CDR technologies (particularly biomass), which are large enough in scale to have a global climate impact, will incur significant development in control of the land surface as an inevitable consequence. To achieve better quality information, it will therefore be necessary to undertake total life-cycle assessment for each technology alone, and then combinations of proposed technological changes. In a natural context these should include interactions with temperature, atmospheric CO_2 concentration, albedo change, and hydrological demands. The utilisation of biochar is a good example, where complex feedbacks and interactions can be expected.[12] In an engineering or human context, the life-cycle assessment needs to consider the engineering construction; supply chain of equipment and materials; sources of energy to undertake the processes; changes in land use; induced changes of land and water use; fugitive emissions and process emissions and induced emissions *e.g.* in soil carbon;[13–16] transportation and its energy sources; utilisation of minor products; as well as cleanup remediation and waste disposal of sites.[15,16] Impacts on society, or on construction, have a time dimension and may result in a "period of payback" to replace emissions during the setup period. Ultimately, agreed frameworks for the scientific analysis of life-cycle and climate in CDR actions will need to be agreed. Moving towards operations, financial and actuarial accountancy, as well as carbon budgeting, will need to be transparent in the life-cycle analysis.[14]

6 Biomass Availability and Sustainability

Biomass is one of the most contested accessible resources. A fundamental tension exists between utilising biomass for CDR *versus* utilisation of biomass for habitat preservation, or utilisation of biomass for food production. Estimation of the supply of sustainable biomass involves many technical, societal, and climatic variables. Availability to within one order of magnitude depends on assumptions of land, fertiliser, water, food demand, climate, biomass type, and technology utilisation developments.[14,17,18] To support ambitious CDR through biomass requires the potential to convert land from agriculture or natural ecosystems rich in carbon, to produce controlled cropping of biomass. This may also involve large inputs of fertiliser or irrigation. The intentional increase of biomass for CDR will often change albedo, for example an increase of boreal forest will reduce winter albedo if snow sheds from dark leaf trees. Integrated modelling will recognise these limits to helpful impacts on climate. The most fundamental human choice of land-use and biomass in the near future will be preservation of habitat *versus* efficient food production *versus* continued and increasing meat consumption.[19] Inspection of the figures cited by proponents suggests large estimates for land biomass potential, and assumes that planned conversion is undertaken and all biomass is used for CDR. Caution is needed because, although some biomass effects can be synergistic, *e.g.* waste derived from crops or forestry, in general biomass uses are exclusive, not additive for CDR. There is also no requirement that a single global approach is undertaken. Biomass utilisation in particular, is likely to exploit optimal regional choices and appropriate technologies.

7 Carbon Dioxide Storage Availability

Options for long-term storage of carbon, $>10\,000$ years, all have geological parameters. These timescales are required to reduce carbon pressures on global change on a long-term or "permanent" timescale. Many of the short-term storage options require continual maintenance and recharging with carbon and are open to variations in societal input, wealth, or extreme climate change. Typical examples could be reforested or afforested regions available for felling, vulnerable to drought, or disappearance due to forest fire. In techno-economic assessments, these generic risks to short-term storage seldom appear to be recognised, in contrast to the durable reliability of long-term geological storage options.

Utilisation of CO_2 is frequently proposed by industrial or business protagonists. However this is fraught with difficulty, when considering the relevant timescale. Utilisation of CO_2 to enhance growth of food (tomatoes and greenhouses), or to augment drinks (carbonated water) indeed has a commercial value, but the storage is fleeting in time, lasting only days or months. Using CO_2 as a feedstock can have a financial value for manufacture of high-value fine chemicals (such as pharmaceuticals), or bulk chemicals (such as carbonate, urea, methane, methanol, formic acid, or liquid fuels).[20] However, if the product is combusted, then that carbon is inevitably released into the atmosphere. Long timescale retention is difficult to justify. It is also clear that the global demand for most of these products is tiny, in comparison to the tonnages of CO_2 produced during power generation.[21,22] Utilisation can, therefore, contribute to the cash flow of early projects, in terms of the present markets, but utilisation is not a long-term storage method or money earner.

Storage of immense CO_2 tonnages can be contemplated below the ground in five types of geological settings. Firstly, by injection into, and reaction with, continental scale basalts – for example the Deccan lavas of India, the Columbia River, USA, and in Iceland.[23,24] Pilot injection tests have been undertaken,[24] but it remains unclear how the majority of rock will be put into contact with injected CO_2. Secondly, utilisation of CO_2 to improve oil recovery has been proven in the USA and several other countries. Commercial methods of CO_2 circulation promote dissolution into remaining oil, or into deep groundwater, making this one of the most secure storage methods, and a method which could find utility as a transient method of generating government taxes, or could be hypothecated to finance capture and transport infrastructure. Set against that, however, is the ugly fact that commercial interests utilise CO_2 to produce additional oil. The carbon balance cannot be reconciled, unless injection to promote CO_2 storage is deliberatively planned, or unless CO_2 injection continues for about the same number of years as oil was produced, beyond the final oil production date. Thirdly, the re-use of depleted methane reservoirs has been proven to securely contain buoyant fluid to recharge limits up to the original natural fluid pressure within the reservoir at the time of discovery. Fourthly, there is the technical possibility of CO_2 injection into sediments of the deep seabed. An optimal zone exists in which liquid CO_2 is denser than the overlying waters, yet less dense than underlying brines.[25,26] This offers potentially immense volumes for CO_2 storage into a failsafe density trapped mechanism. However, the logistics of transport from CO_2 sites to the deep offshore, and injection followed by monitoring, are financially challenging. The fifth approach is the direct injection of liquid CO_2 into regionally widespread saline aquifer formations. These are routinely assessed to be abundant within the oilfield areas currently under investigation, and comprise more than 80% of accessible storage volumes of commercial interest. Evaluation of this type of storage, to the required level of commercial certainty, may require hundreds of millions of pounds to drill test boreholes and produce fluids, for each regional saline formation hosting one gigatonne CO_2 or more. There is good potential to reduce that exploration cost by using legacy boreholes, or existing holes, tracked sideways to inject test CO_2 at more than one horizon within the layer cake stratigraphy.

The conventional assessment of CO_2 storage, seeking sites in saline aquifer formations, still has large uncertainties.[27,28] Improving the site-specific estimates of storage available can be achieved by a combination of improved subsurface geological information, such as seismic reflection surveys or direct drilling. Good examples of progressively improving evaluations of storage are provided by the North American Atlas,[29] the Norwegian Petroleum Directorate surveys,[30] or the UK databases.[31,32]

There are also many options for engineered enhancement of CO_2 storage in single saline formations, such as large-scale production of deep saline water, to be disposed at the surface. This creates "voidage space' in the deep subsurface which can easily raise CO_2 storage utilisation efficiency from 2% (the amount dissolved in saline water) by a factor of four.[33] Nevertheless it is clear that sedimentary basins, which could have characteristics for geological storage of CO_2, are not uniformly distributed around the Earth.[28] The practical problems arise in detailed matching of CO_2 sources with the nearest storage sink, *i.e.* connecting capture to saline aquifer. Established methods involve construction of overland pipelines, which can be tens or even hundreds of kilometres in length.[27,28,34] CO_2 shipping is also possible using tankers converted from liquefied petroleum gas. All these add expense and complexity, as well as difficulties of public permission. In principle, it is therefore a sensible strategy to relocate surface CO_2 capture facilities above storage sites. This results in many fewer potential sites, for example when using air capture, where the appropriate meteorology and climate lie above the appropriate geological storage.

In simple arithmetic terms the known commercial reserves of fossil hydrocarbon today can be approximately balanced by the estimated global quantity of storage resource. However, if, as has been the historical precedent, the much larger resources of fossil hydrocarbon are gradually converted into commercial reserves, then the amount of known storage on land requires major engineering to improve subsurface storage volumes. Even then, our estimates are that fossil hydrocarbon resources greatly exceed the summation of conventional geological CO_2 storage.

The performance of deep geological storage during long timescales is often the subject of debate. There is an element of dual standards in such discussions when, for example, a reforestation proposition is regarded as more secure carbon storage them deep injection of CO_2. We suggest that forests can be significantly liable to 100% loss of carbon by fire or drought and that is regarded as acceptable. Whereas by contrast geologically stored CO_2 is unlikely to leak, but monitoring technologies struggle to detect leakage rates of 1% per 1000 years, and this is sometimes regarded as unacceptable. We propose that the possibility of modest rates of geological leakage still retain many benefits of carbon reduction which assist climate mitigation in the medium term.[35,36] There is clearly a balance to be struck between tonnages of CO_2 stored for maximum security but with much less available storage capacity, *versus* strategies where storage capacity is maximised whilst accepting a statistical probability of slower rates of long-term leakage.

Figure 2 illustrates the advantages and limitations of the various carbon dioxide storage methods (magnifying details of CDR from Figure 1). The X axis estimates the range of possibility, the Y axis estimates the feasibility expressed as cost per tonne of CO_2. Darker colour shades indicate greater maturity of the technology, for example afforestation is well understood relative to capture. Estimates are gained from the various sources cited in Table 1. As explained previously, many figures are uncertain both in terms of cost and potential tonnage per year. Nevertheless, this diagram provides a visual estimate of the most important actions, which could have greatest potential impact in capturing and storing carbon stock. It is clear that the technologies with the greatest claims are biochar, biomass with carbon storage, and air capture. Note that there may be resource conflicts between biochar and biomass with CCS, as published estimates for maximum deployment assume that each is the sole dominant technology. The conclusion from this is that much greater certainty of costs is needed for all three technologies, and much better estimation of biomass resource availability is needed in the context of food production, and maintaining sustainable

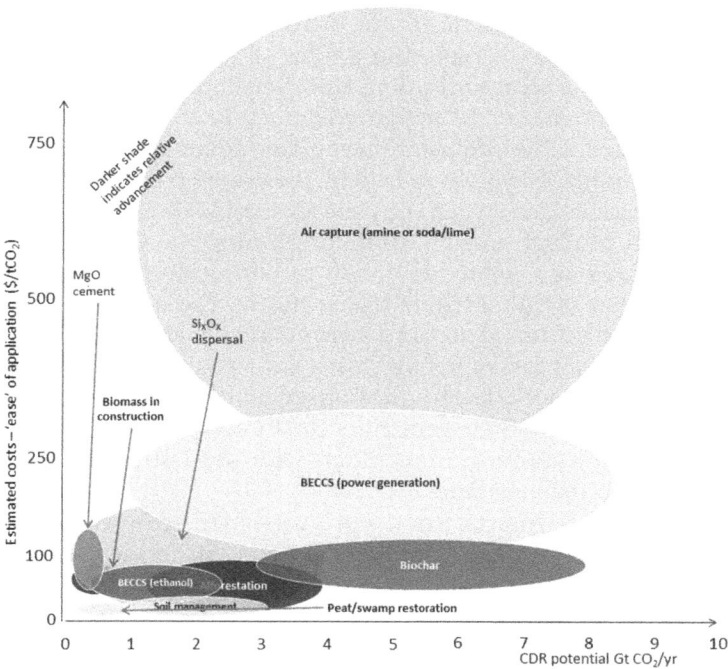

Figure 2 Diagram showing the feasibility, the CO_2 removal potential and maturity of technology for various CO_2 storage methods.

terrestrial ecosystems immune from multi-decade timescale forest harvesting, or forest gardening.

8 Summary of Carbon Storage Methods

8.1 Increased Terrestrial Biomass: Afforestation

Essential features: planting more trees is easy to understand in principle, but rather more complex to calculate in terms of its benefits. There is a distinction between afforestation and reforestation. Afforestation is where trees are replaced in regions where they have been absent for greater than 50 years, and can also conceptually include increasing the carrying capacity of a landscape for unharvested terrestrial biomass. The concept of reforestation, on the other hand, is where forest loss occurring within the past 50 years is replaced.[37] It is also important to make a distinction between tropical and temperate forest systems, since tropical biomass grows rapidly, with a short replacement timescale, and with minimal albedo effects. By contrast temperate forest systems grow 5 to 10 times more slowly,[43] and can actually decrease albedo by darkening the landscape especially during northern winter when reflective snow can be replaced by an absorbing surface of dark coniferous needles.[42]

Potential impact: estimating the carbon storage impact of forestation has great uncertainty. Optimistic maximum estimates speculate on the regrowth of all deforested regions,[11] replacing a total of 180 $+/-$ 80 gigatonnes of carbon, including soil recarbonisation. However, competition for land use in agriculture means that a realistic figure is much lower.[19] A global potential maximum for sustainable afforestation could be about 1.5–3.0 Gt CO_2 yr^{-1}.[38] Deforestation produced by human activity is estimated to emit 2–4.5 Gt CO_2 yr^{-1}.[39] Complications arise when the payback time is considered.[40,41] Felling of forests could produce a rapid emission of biogenic CO_2 if the wood is combusted, as well as a reduction of soil carbon during harvesting and re-planting.[41] Carbon is only actively stored during the maximum growth of new forest, typically from year 10 to year 40 after planting. Subsequent to that timescale, forests can achieve a steady state position with much slower sequestering of atmospheric carbon. Consequently, to utilise reforestation requires continual active management of the forest carbon stock on a global scale. That intensification of use conflicts with most strategies for conservation and habitat management.

Costs: financing the forest carbon stock varies greatly in cost, depending on competition with local demand in agriculture for fertile cropland, from $20 to >$100 t^{-1} CO_2. There is also uncertainty in the measurement of stocked carbon – terrestrial methods are slow and labour-intensive, whereas satellite radar methods are much more rapid but are only just emerging as a method. Forests also need to be maintained in very different ways in different settings, and the payment for this is unclear, other than by commercial extractive forestry. Although programmes are being developed such as Reducing Emissions from Deforestation and Forest Degradation (REDD) which aims to offer incentives for developing countries to reduce emissions from forested lands and invest in low-carbon paths to sustainable development.

Security, effort, adverse effects: forestry is well understood in its fundamentals. There is a clear risk from local competition in land use – a problem which needs to be guaranteed through many decades. There can also be catastrophic risk of tree death from drought or from forest fire. In addition, maintaining maximum rates of carbon stocking requires rotational felling after only a few decades, removing the possibility of a long-term stable ecosystem. Adverse effects of forestry include water consumption, albedo change, and the potential to alter regional cloud patterns and rainfall.[19,38] Stocking with rapidly growing non-native species can introduce fungal or pest infection. Cultural changes can also be induced, such as the use of fuel wood, hunting and foraging for food, and the care for valued wildlife.

8.2 Increased Soil Biomass: Biochar

Essential features: 'biochar' is a term applied to charcoal produced through low-temperature pyrolysis, intended for utilisation in a soil ecosystem. The placing of char intends to increase the lifetime of biomass carbon within actively managed soil profiles. This can be applied to agriculture or forestry, at local farm or small industrial plant scale, and can use excess biomass derived from many types of feedstock. A range of pyrolysis conditions and time durations at 300°C to 900 °C, span the range from high retention of volatile organics in the char through to complete gasification in a process where oxygen for combustion is incomplete. Co-products include impure bio oil, and flammable syngas which can be used to power the pyrolysis equipment.[12,44,45]

Potential impact: using current land behaviour estimates, and applying biochar globally within all applications and identified niches places carbon stock management with biochar at around 3.5 Gt CO_2 yr^{-1}, with a cumulative total of 500 Gt CO_2 sequestered during 100 years.[12] An optimistic calculation, where biochar is applied to all agricultural grassland areas, derives a maximum global accumulation of 1500 Gt CO_2 (400 Gt C) during 100 years.[12]

Cost: financing biochar depends on multiple factors: predominantly feedstock, transportation, and labour costs. In Western countries, if biomass costs are $<$\$100 t^{-1}, the least cost applications are \$40 t^{-1} CO_2, ranging up to \$150 t^{-1} CO_2.[45,46] Biochar yield can be estimated as upwards of 25% of the mass from dry feedstock, producing a sequestration of 0.46 t CO_2 per dry tonne of biomass feedstock.[47] In common with many "negative emission technologies", biochar cannot currently compete against the rival proposition of simply burning all the biomass, and thus returning CO_2 rapidly to the atmosphere. This type of problem requires government action to create payments which reward long-term carbon storage. Pyrolysis equipment for multiple small-scale applications is available now, although there is substantial scope for improved control and specification of pyrolysis.[45]

Security and effort: there are several well-publicised examples of Amazonian "terra preta" biochar, which have lifespans of several hundred years in the soil. Matching biochar type to soil type is a topic of active research, and establishing residence times of carbon stock between decades and several centuries seem very probable. Active management is required during initial application, with minimal maintenance thereafter. Routine large area verifying of biochar stock in soil is, as yet, undeveloped, although proxy methods of remote sensing for carbon in soil are under investigation, and manual sampling and analysis is established but slow.

Adverse effects: biochar costs are locally specific, affected by factors such as biomass feedstock prices; costs of collection and handling; transportation energy used; and effective pyrolysis control. Once emplaced into soil, large biochar fragments could be harvested for use as fuel, thereby rendering any benefits void. There are no known negative health impacts from carbonised dust, or mobilisation of volatiles in the soil. Exposure of biochar at the soil surface will reduce albedo, potentially by 13–22% at steady state in the year after application.[48,49] If biochar is applied annually, then large albedo reduction could continue before reaching steady-state.

8.3 Biomass Energy with Carbon Capture and Storage (BECCS)

Essential features: biomass can be combusted to produce heat and power utilising non-fossil carbon. Biomass is currently co-fired in small quantities with fossil coal or lignite, providing about 3% energy input (biomass ranges between 30–80% of the energy density of steam coal.[50] About 1.55% of global electricity was from biomass in 2010.[50] If capture and storage is undertaken on the combined flue gas, then that results in net extraction of carbon from the ambient atmosphere into deep geological burial.[10] Offset against that needs to be a full life-cycle analysis of energy and fertilizer used in planting, maintenance, harvesting and transport, which reduce many of the claimed benefits. Co-firing also introduces problems, such as greater variation of impurities, less concentrated CO_2 in flue gas, variable burn in oxyfiring[51] and disposal of fibres and tar during gasification.

Similar extraction by biomass indirectly from the atmosphere can result if the pure CO_2 waste stream is captured from ethanol production by fermentation, and disposed into a deep geological reservoir. The leading example of this process is the plant of ADM at Decatur, Illinois, USA.[52] A simple calculation of project cost *versus* tonnage stored, shows that this produces a cost of around $60–70 t^{-1} CO_2. About 85 billion litres of ethanol are produced annually worldwide, which co-produces 68 Mt CO_2.

Finally, bio-methane can be produced by anaerobic digestion, and is added into the gas grid or co-fired in gas power plant as carbon-neutral fuel – where CCS may, eventually, be undertaken.

Potential impact and security: there are three main problems which reduce deployment of bio-ethanol and biomass power. These are biomass availability and sustainability, availability of CO_2 storage, and conversion of legacy infrastructure.[53] The potential of BECCS is estimated at 2.5–10 Gt CO_2 yr^{-1}. The larger estimates include considerable conversion of agricultural land to production of feedstock.[54] The security of long-term storage is identical to that for CCS, *i.e.* permanent CO_2 removal in climate terms.

Costs of operation: estimates vary greatly, because of different assumptions of the value of electricity, or transport fuel, combined with potential cost reductions through improved CCS. Storing CO_2 as a by-product of ethanol is expected to represent the lowest cost, like at the Decatur plant,[52,54] less than the cost of capturing CO_2 from co-fired biomass. Estimates for BECCS at \$100 t^{-1} CO_2 are considered too optimistic,[55] because they are less than the projected costs of CCS with fossil fuel.[56]

Effort needed: BECCS is a continual and intensive resource process, to gain reliable and regular feedstock supply. As with CCS, the capture process and CO_2 compression before transport devour a large proportion of the stated energy input. Regional transport networks from capture to storage would ideally augment or inherit conventional CCS pipelines.

Adverse effects: as with any biomass technology, to have a large impact, this will confront the competition for land-use between energy, food and water. Even though conventional CCS has societal acceptance in most parts of Europe, the industrial aspects of BECCS, and association with coal fueled power plant may reduce its acceptability.

Figure 3 compares commercially available fossil fuel reserves (gigatonnes (Gt) CO_2) with the potential options for CO_2 storage (Gt CO_2) which are ranked by timescales of isolation from the atmosphere. To enable recovery of the planetary climate system, timescales of at least 10 000 years are required, shown by climate models and by the geological record of recovery from past high CO_2 excursions. Differential shading of columns indicates the high and low estimates. Only if estimates of fossil fuel commercial reserves are low is there any possibility to balance with the highest estimates of the CO_2 storage available. That is unlikely. Furthermore, commercially cited reserves of fossil hydrocarbon are expected to be about 10 to 100 times less than the commercially ill-defined natural resource of fossil carbon available.

8.4 Biomass Burial, Carbon Dioxide Use and Algal Carbon Dioxide Capture

Essential features: a simple method of carbon stock storage, is to bury biomass. Normal human operations introduce waste biomass organic material, such as crop waste, manure, or compost, into agricultural land. This could potentially reach 2 Gt CO_2 yr^{-1},[57] however, carbon residence time is extremely short, only years. In agriculture, changed management practices may enable additional carbon to be stored in soil, for example by no till ploughing which reduces carbon loss through oxidation.[57]

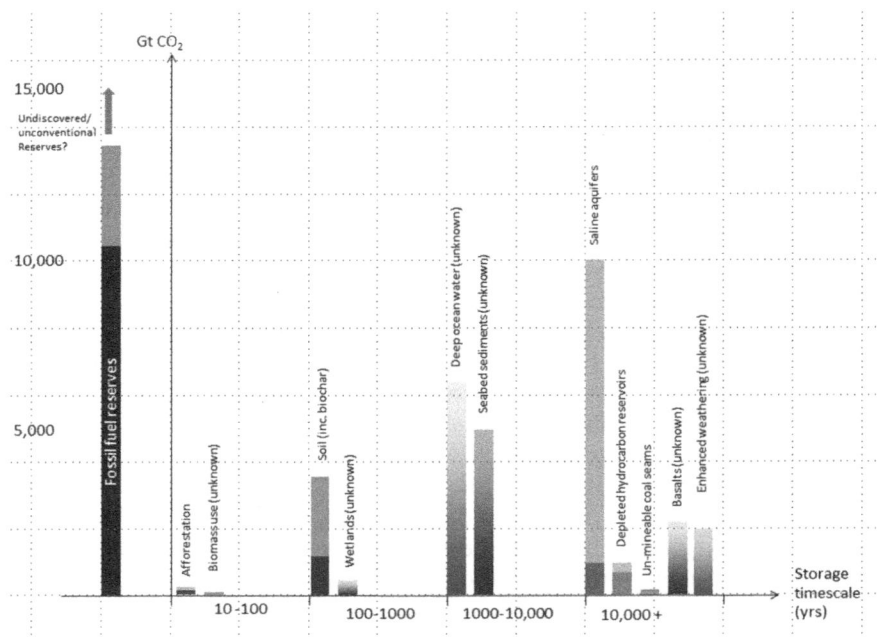

Figure 3 Diagram showing a comparison of commercially available fossil fuel reserves (gigatonnes (Gt) CO_2) and the potential options for CO_2 storage (Gt
CO_2) which are ranked by timescales of isolation from the atmosphere.

As a scale-up of this principle it is proposed that biomass could be buried
in the deep ocean,[58] where residence time may be hundreds to thousands of
years. Especially slow decay may occur where some parts of the Antarctic
Ocean may be isolated, for thousands of years, from wood decay
organisms.[59]

Another step to industrialising the use of biomass to capture CO_2 and fix
carbon is the active development of algae in bioreactors and the adaptation
of enzymes as organic catalysts for air capture. The energetic feasibility,
costs, and size scale of impact remain poorly known.[60–62]

The use of biomass is well established in construction, predominantly
with timber use in houses, but also with potential for installation utilising
straw. This will decrease the use of cement, offsetting emissions during its
manufacture, but the overall potential is fairly small.[63]

8.5 Direct Air Capture

Essential features: the engineering of direct air capture carbon dioxide reduction is applied to the extraction of CO_2 from ambient atmosphere, with
concentration and permanent storage of the captured CO_2. A large field of
aspirational technology inventors exists, with the leading 10 promoted
through media such as the Virgin Earth challenge,[64] competing for a

$25 million prize. Because of the immensely large commercial potential, precise details of the technologies are hard to obtain.

Processes: widely publicised approaches are: (i) adsorption onto solids, or (ii) absorption into high alkalinity solutions. All methods face three fundamental challenges: (i) overcoming the large thermodynamic barrier (theoretically 500 MJ tonne^{-1} CO_2). This makes the proposed reactions almost impossible to conceive in isolation. The energy barrier is high because of the low CO_2 concentration in air (0.04%);[65] (ii) supplying and sustaining sufficient airflow and immense air volumes through the system using minimal energy to contact with the active surfaces; and (iii) supplying energy to regenerate the active reagents, to compress the CO_2 before pipeline transport and to inject to deep storage. Examples of three of the leading contenders are methods proposed by Global Thermostat, Klaus Lackner and David Keith. Global Thermostat seeks to use amine coated cellular solids to absorb CO_2 directly from air. Amine is regenerated using low-grade process heat associated with power plants or refineries.[66] Thus, successful operation of this technology is intimately dependent on the fossil fuel combustion it seeks to offset. Alternatively, the method proposed by Klaus Lackner[67] involves nonproprietary amine based resins that can capture CO_2 from ambient air movement, and CO_2 is released by hydration under reduced pressure.[68,69] In spite of detailed laboratory measurements, the energetics of these proposals remain intensely contested.[65,72,73] The third method, proposed by David Keith,[70] adapts a well-established circular chemistry method from paper manufacture. Carbon dioxide is absorbed under forced air fan flow by contacting potassium hydroxide which is converted to potassium carbonate. The potassium is regenerated to hydroxide, by reaction with sodium hydroxide, and the resulting sodium carbonate is regenerated by a calcination reaction heating to 900 °C by burning additional methane to release the dissolved CO_2. All emissions, including CO_2 gas heating, are captured within the system.[71]

Costs and feasibility: all these methods are currently in their experimental or small pilot stage. Carbon engineering is progressing most rapidly. The estimation of cost varies greatly, and is intensely contested. Developers claim anything between $90 per tonne CO_2, and $200 per tonne for pilot plant, with expectation of cost reductions at a larger scale.[73] External estimates are always much higher, based on costs of equipment or on fundamentals of thermodynamics; these range from $600 per tonne minimum,[72] to upwards of $1000 per tonne CO_2.[73]

Deployment, adverse effects: the scale of deployment remains unknown. If air capture can work reliably and economically, it is possible that one to 2 gigatonnes CO_2 per year could be extracted. It is equally possible that such development will drive these technologies towards flue gas as a rich source of CO_2 which may restrict air capture into niche markets, including (ironically) the potentially profitable proposition to make CO_2 on site for enhanced oil recovery. Envisaging scale up to impact on atmospheric CO_2 concentrations requires a large leap of faith.[72] Calculation suggests that offsetting CO_2 emissions of the UK by air capture will require tens of thousands of installations around the UK, each sized similarly to the cooling tower of a conventional coal fuelled power plant. The visual impacts will make onshore wind power seem benign.

Air capture approaches could also be applied to other greenhouse gases, for example it is claimed that methane might be commercially extracted from ambient air.[74]

8.6   Silicate Weathering

Essential features: in the natural climate system, weathering of silicates by reaction with atmospheric CO_2 is an important drawdown to maintain climate equilibrium. In the perturbed human system at the present day the natural rate of weathering is about 1% of the rate of CO_2 emission. Enhancement of weathering can be achieved by increasing the surface area of reactive minerals in contact with the atmosphere or ocean. This could be achieved by grinding minerals to smaller particle size and then distributing on land or at sea – or by adding reactive minerals to naturally abrasive settings such as beaches. Two main styles are usually considered, the dispersal of silicate olivine across land,[75,76] or replacement of carbonates in cement by using Mg oxides or silicates or fly ash.[77]

Engineering feasibility: the relevant reactive minerals are abundant in some parts of Earth's surface, for example olivine can be quarried from large parts of Oman. Basic chemical reactions suggests that one tonne of olivine is required to sequester 1 tonne of CO_2. Consequently enhanced weathering methods all require extraction and crushing of rock volumes which are similar in tonnage to those involved in the extraction of fossil fuels. To replace all carbon emitted from cement production, would remove about 5% of human emissions, about 2.5 gigatonnes CO_2 per year, but only 20% of this is considered to be accessible through commercially feasible processes.[78]

Costs: extraction of bulk materials is a low-cost operation, typically $3–$7 per tonne. Landscape remediation, transportation costs, and distribution costs also need to be considered and could double that estimate. Nonetheless this appears to be a low-cost option. Although some reactions look attractive on paper, for example replacing carbonate in cement with magnesium silicates removes 0.6 t CO_2 per tonne cement,[77] the immense tonnages of magnesium silicate may only be readily accessible in some parts of China for example – producing much larger costs of transportation.

Adverse effects: although the fundamental reaction of olivine weathering may absorb CO_2, the reaction products run off ultimately to the ocean to create alkaline compounds which, under large scale rapid deployment, would lead to unknown changes of ocean chemistry. The creation of an immense quarrying industry at the land surface, to balance subsurface extraction of coal oil and gas, would utilise land area perhaps four times the extent of coal strip mining.

8.7 Chemical Feedstock

Essential features: there is a burgeoning industry in developing green chemistry which can utilise CO_2 as a reagent.[79] The products include various categories of compounds: fuels, polymers, carbonates, carboxylates, as well as direct bulk applications such as solvents in enhanced recovery, or solvents or growth enhancers in the food industry. The generic problem, of course, is that CO_2 is a molecule of low chemical potential energy, with a Gibbs free energy value that is very negative ($\Delta G = -400 \text{ kJ mol}^{-1}$), and is difficult to react. Current approaches include: using catalytic methods which utilise hydrogen; harvesting wasted or sustainable energy to drive the reactions; or coupling CO_2 consuming reactions with very exothermic parallel reactions.

Costs feasibility: costs are poorly constrained, and depend closely on the particular reaction undertaken and its method of operation. As with many chemical processes it is perfectly possible to obtain a suite of minor reaction products, the problem being to control side reactions and most of all to increase yield in an energy effective process.

Security of retention and effort: many of the compounds produced have only a short lifespan of human usage until they are recycled into the atmosphere. For example, urea as a bulk chemical can be used in fertiliser, but is then released to atmosphere within months or years – much too short to be of significant impact as a carbon storage method. Consequently the manufacture of chemicals would need to be a continuous effort in drawing down CO_2. A real problem though, is that of scale. Compared to the quantity of CO_2 released from human combustion of fossil fuels, the CO_2 tonnage utilised for chemical products is only a few percent. As a very optimistic example, in 2012 the entire EU27 demand for CO_2 as a feedstock of 235 M tonnes is exceeded by CO_2 output from industrial processes alone (300 M tonnes CO_2).[79] This does not consider the emissions from power and heat generation, so that CO_2 utilisation may at most use 10–20% of emissions. Although chemicals containing CO_2 can provide a high-value income stream to a capture project, the market is easily saturated in commercial terms – especially locally. One example is salicylic acid.[80] Just 60 tonnes day^{-1} of CO_2 from a gas power plant slipstream, can manufacture the global demand. Thus, CO_2 as a chemical feedstock has potential uses in engaging with industries and communities, but it needs a market to be created for its products and/or to find an energetically suitable way of re-making fuels which recycle carbon from emitted CO_2.

8.8 Carbon Dioxide for Enhanced Oil Recovery (CO2-EOR)

Essential features: in North America and China, the processes of carbon capture and storage are usually discussed as "carbon capture utilisation and storage". This utilisation proposes that a large part of CO_2 captured from power plant and industry should be transported for injection into partially depleted hydrocarbon fields. Carbon dioxide enhanced oil recovery

(CO2-EOR) has been undertaken in the USA since the SACROC Texas project commenced in 1972. At cool temperatures and low pressures, the CO_2 fluid can act to re-pressurise the hydrocarbon field and physically drive out additional hydrocarbon production. The CO_2 fluid is even more effective at higher temperatures and elevated pressures, however, as it becomes fully miscible with the hydrocarbons and acts to decrease surface tension. This can enable production of an additional 5 to 20% of the original oil in place. If commercial restraints did not feature, and extended timescales are available, then it is possible in principle that CO_2 injection could move towards production of 100% of the original oil in place. This process is often advocated as a beneficial, sustainable, and economically valuable utilisation of CO_2, which results in net storage in depleted hydrocarbon fields. Between 44 and 84 billion barrels of additional oil are calculated to be producible in the onshore USA alone using CO2-EOR.[79] However, there seems to be a clear arithmetic bear-trap in the life-cycle analysis, being that large quantities of additional hydrocarbon are produced which could counterbalance the CO_2 stored. Convention in the USA has it that any produced hydrocarbon "does not count" as an emission for climate budgeting purposes of a project. Whereas in Europe, the convention is that any production of hydrocarbon counts as an emission for climate purposes, even if not counted for trading scheme purposes until combusted. In established CO2-EOR in the USA, about 160–300 kg of CO_2 are used to produce 1 barrel of oil, which emits 430 kg CO2 on combustion.[80] Clearly more carbon is produced than is stored. However, the picture is complicated by the economic fact that CO_2 to undertake EOR in the USA costs money, paid by the oilfield operators. Therefore a commercially profitable operation requires minimisation of CO_2 purchase and use. The market system, as it currently exists, does not place a levy on carbon emitted or repay value to carbon recovered and stored. By contrast using CO_2 for the purposes of sequestration may promote additional CO_2 storage, but only if the CO_2 has a disposal value attached. It is possible to re-design CO_2 injection for EOR, such that an overall storage of carbon is the result.[81] A combination of credit for CO_2 disposal plus enforcing environmental legislation will be required to ensure that CO2-EOR is environmentally beneficial, rather than adding even more emissions to an acute problem.

8.9 Deep Sea Sediments

Essential features: although there is much discussion of enhancing CO_2 accumulation in shallow ocean water, and dispersal into deep ocean water, there has been little investigation of utilising vast areas of sediments along the continental margins and the deep ocean. One theoretical possibility is to inject CO_2 into sediments deeper than 3000 metres water depth and underlying several hundred metres of sediment column. This combines the useful effects of geological storage, combined with large volumes of ocean storage, and suitable geochemistry.[25] When CO_2 is injected into the ocean deeper than 3000 m it sinks due to density. The overlying pore fluid forms a less dense cap to prevent buoyancy migration. This zone is also below the pressure temperature conditions for CO_2 hydrate formation, ensuring a second category of seal. Carbon dioxide that is injected beneath this zone gradually dissolves into ambient pore water and becomes denser, and sinks.[25,26]

Cost, feasibility: no costs are available for this method. Costs are likely to be high. CO_2 needs to be transported offshore by tanker, because injection sites will need to be laterally extensive. Advanced drill ships will be needed to hold the drill riser static whilst injection occurs. Physical conditions of the target sediments may not be helpful, due to small grain size and intrinsically poor permeability, leading to bad injection rates.

Tonnage: as suitable sediments are widespread on all Atlantic type continental margins, the potential storage volumes are extremely large – greater than 2000 years of current USA CO_2 production.

9 Discussion

It is clear that many and diverse opportunities for storing carbon exist (see Table 1). However the practical difficulties of re-capturing feral fossil carbon are immense. The experience of CCS is salutary.[82,83] Since the mid 1990's the ambition to develop CCS has been expressed by the governments of many industrial countries. However, even working within the present electricity generation system, the conversion from "established" to "clean" methods of operating has stalled, to run at only 10% of the build rate required to achieve climate sustainability. There are regulations to draft; territorial claims to make; regulators to create; licensing allocations to decide; pipelines to convert or build; expensive injection boreholes to drill; and monitoring and verification to enact. And that is all before tackling "who pays". So, by the mid 2010's, not a single UK project has been built, and only a small handful are operating globally. Slow, tortuous progress is normal, even for a technology which is conceptually easy to understand and is highly favoured by governments. What is the chance, then, that innovations such as biochar, or soil recarbonisation, will progress any faster? The message from other innovative technologies is clear: establishing acceptability can take 15 years from university to industry then scale-up of deployment usually takes a long time – several decades. Exceptions exist and are usually framed as responses to crises, emergencies, or wartime.

A fundamental blockage exists with money. Who pays for the common good is not politically a vote winning formula. There are different ways of analyzing this problem. On one hand, the UK claims only 2% of global emissions (although the embedded emissions in imports are roughly double that, and the UK stock market influences an additional 25% of global GHG production). On the other hand, the UK, being first to industrialise, has accumulated a historic debt of carbon emission, which places UK citizens

R. Stuart Haszeldine and Vivian Scott

Table 1 Summary of CDR methods.

CDR method	Description	Estimated potential (Gt CO₂ yr⁻¹)	Estimated cumulative potential (Gt CO₂)	Residence time of removed CO₂ (years)	Vulnerability of removed CO₂[a]	Abatement measurement
Biomass based processes						
Afforestation	Increasing forest cover	1.5–3	300–500 (ref. 69)	$10–10^2$	High	Complex but in development for *e.g.* REDD
Wetland enhancement	Increasing peat land C uptake	<0.5	unknown	$10–10^3$	High	Difficult: C fraction retention and time period subject to many factors
Biochar	Charred biomass with high stable C fraction dug into soils or buried	<3.5	500	$10–10^3$	Med	Difficult: C fraction retention and time period subject to many factors
BECCS – ethanol	Biomass fermentation with CCS	<2 (unknown)	unknown	$>10^5$	Low (if geological)	Measurable
BECCS – electricity generation	Biomass burning with CCS	2.5–5	350 (ref. 70)	$>10^5$	Low (if geological)	Measureable
Biomass burial	Burial of waste or purpose biomass	<2	unknown	$1–10^2$	Med	Difficult: C fraction retention and time period subject to many factors
Biomass use	Biomass in construction	<1	unknown	$10–10^2$	Med	Measureable
Algae	Air or flue-gas capture with CO2 storage or biofuel creation	unknown	unknown	$>10^5$ (if geological storage)	Low (if geological)	Measureable
Chemical processes						
Direct Air Capture – solid adsorption	Artificial trees	unknown	unknown	$>10^5$ (if geological storage)	Low (if geological)	Measureable
Direct Air Capture – alkaline solutions	Adsorption by sodium hydroxide	unknown	unknown	$>10^5$ (if geological storage)	Low (if geological)	Measureable

Estimated cost $ t^{-1}$ CO_2[b]	Resource requirement[c]	Primary limitations	Control-reversibility: low, med, high[d]	Deployment speed (years)	Technology readiness level (1-9)[e]	Side effects and impacts	Research challenges
$20–100	LA***** NR*** MR* EI* LS**	Land-use competition	High-high	$>10^1$	6–7	Albedo change, hydrological effect, societal landscape use change	
$20 +	LA*** NR*** MR* EI* LS**	Land-use competition and water	High-high	$>10^1$	5	Land-use competition, possible CH_4 emissions, Ecosystem benefits.	
$30–40	LA**** NR***** MR** EI* LS***	Biomass availability	High-low	$>10^1$	5	Land-use competition, Improved soil fertility, possible carcinogenic dust, albedo change	
$25 + (unknown)	LA**** NR***** MR*** EI** LS****	Biomass availability	High-med	$>10^1$	6	Land use competition, water demand	
$100–200	LA*** NR***** MR*** EI* LS****	Biomass availability, CO_2 storage availability	High-med	$>10^1$	4–5	Land use competition, water demand	
Unknown (low)	LA*** NR***** MR* EI* LS**	Biomass availability, suitable land	High-unknown	10^1	4–5	Local environmental change	
Unknown (low)	LA**** NR***** MR* EI* LS***	Demand	High-med	10^1	7–9	Societal	
unknown	LA*** NR*** MR*** EI** LS****	Nutrient	High-high	unknown	2–5	unknown	
$100–500 + (unknown)	LA*** NR* MR**** EI**** LS****	Land area, CO_2 storage availability, energy	High-high	unknown	2–4	Land-use	
$100–500 + (unknown)	LA*** NR* MR**** EI***** LS****	Land area, CO_2 storage availability, energy	High-high	unknown	2–4	Land-use	

Table 1   Continued.

CDR method	Description	Estimated potential (Gt CO$_2$ yr^{-1})	Estimated cumulative potential (Gt CO$_2$)	Residence time of removed CO$_2$ (years)	Vulnerability of removed CO$_2$[a]	Abatement measurement
Accelerated weathering	Pulverised silicate dispersal	0.1–3.5	400	$>10^5$	Low	Complex
Magnesium oxide cement	Negative emissions cement	<0.5	unknown	10^2–10^3	Med	Measurable

[a]The vulnerability of the removed CO$_2$ is ranked: high, medium and low, with respect to possible future climatic, environmental or societal impacts. Afforestation is ranked has 'high' as it may be subject to drought, disease or changed societal demand. Geological CO$_2$ (appropriately sealed) by contrast is ranked 'low'.
[b]Cost estimates are highly subjective and are to a large extent based on current costs of resources and materials that may not remain valid.
[c]Resource demands are qualitatively assessed on a scale of * to ***** (highest) according to following categories: LA: land area; NR: natural resource demand (*e.g.* water, biomass, mined substance); MR: manufactured resource (*e.g.* steel, synthetic chemicals); EI: energy input (net); and LS: logistical scale (distribution, transportation).

amongst the most polluting global nations per capita – and it could be suggested that the UK should now, immediately, devote several percent GDP into cleaning up its historic legacy of carbon emission on which most of its present wealth is founded. Who pays to scrub CO$_2$ from the common global air is the type of question on which it is particularly difficult to convince sceptical voters.

The climate calculator is clear. To avoid the 2 °C rise in average sea and land temperature predicted by climate modeling requires actions to drastically cut emissions, and then to keep within a budget of total carbon emissions.[5] That budget, of one trillion tonnes of carbon, expires in February 2044 if humans persist in their established behaviour. So a combined approach is needed to enforce a reduction in emissions rates, combined with a firm cap.

So, what happens to the basic arithmetic? We know the total carbon per year emitted. We also know the total commercially exploitable carbon in economic reserves, and in the technically potential resources. And, from the analysis undertaken in this paper, we can estimate the total carbon storage resource. The numbers do not match. Our known resource of combustible carbon is far greater than the immediate storage ability.

That means changes, for example, the development of energy storage batteries which could store energy for a day, or even a week, as well as an improved understanding of how to reduce rebound behaviour where CO$_2$ savings in one sector may sometimes be transferred and expended in a different sector. This leads to the perennial Jevons paradox,[84] which suggests

Estimated cost $ t^{-1} CO_2[b]	Resource requirement[c]	Primary limitations	Control-reversibility: low, med, high[d]	Deployment speed (years)	Technology readiness level (1–9)[e]	Side effects and impacts	Research challenges
$25 + (unknown)	LA*** NR**** MR*** EI*** LS*****	Logistics of application	High-low	unknown	2–3	pH increase of land	
unknown	LA* NR**** MR** EI*** LS***	Demand	High-med	unknown	2–4	Mining of Mg	

[d]Controllability and reversibility are ranked: high, medium and low. Since afforestation can be ceased and felled, it is ranked 'high' both for control and reversibility.
[e]The stage of technology development is assessed on a scale of 1–9: 1 (scientific principle identified) to > 9 (proven and deployed system).[71]

that if energy efficiency increases, then energy consumption also increases. However, there are more specific studies which show that Jevons "rebound" does not always form a paradox, especially if the price, or other rationing, of energy increases at the same time as efficiency increases.

For the global carbon budget, there are several ways to react to the unfortunate mismatch. You can pretend nothing is wrong and carry on as before. You can make encouraging noises, then develop and deploy CCS and CDR components enroute to a real operation.[83] Or, you can mediate a rapid transfer across to low carbon emissions, which ultimately means getting out of fossil carbon utilisation as the main basis for energy supply. The latter would require a fundamental change to the behaviour of industrial societies, where taxing the dis-benefits of fossil carbon for a reduction in its use has never been on the agenda. Nevertheless, the evidence in this analysis suggests that sustainable carbon dioxide reduction or stabilisation can not be achieved by replacing continuing, or historical (from air), emissions of carbon into the ground. Leaving fossil carbon in the ground, unburned, is today a radical statement, but this may yet become the least expensive, and lowest risk, option.

10 Conclusions

1) Many methods of carbon dioxide reduction (CDR) exist (see Table 1). All of these have significant uncertainty in their global potential and poorly constrained costs.

2) All methods of CDR will require global industries to arise and deploy new logistical systems. Previous examples of industrial step changes have taken many years from invention to industrialisation, and several decades to full-scale deployment. CDR will not be rapid.

3) Rational life-cycle analysis of CDR methods has not usually been undertaken, based on carbon, energy, or resources.

4) The different CDR methods are sometimes in conflict, particularly for a finite supply of biomass, where rival uses include food production and long-term ecosystem maintenance.

5) Even though CO_2 storage capacities of all methods are very uncertain, it is clear that CDR alone is very unlikely to achieve net reduction of atmospheric carbon, because CDR cannot balance the projected release rate of fossil carbon, even with all standard mitigation efforts and fuel switching included.

6) Leaving fossil carbon in the ground, unburned, is the best option for humanity to move beyond hydrocarbon for long-term sustainability. To make this transition, new and competitively costed energy sources need to develop, and older carbon emitting activities need to be penalized. Both of those need courage by governments, which can arise from leadership, or from pressure by citizens.

Acknowledgements

Much of this work was prepared during funding by EU-Trace FP7, operated by M Lawrence at IASS Potsdam. RSH is also supported by the Scottish Funding Council (SFC), EPSRC, NERC and ScottishPower. VS is supported by EU-Trace FP–7 and SFC.

References

1. P. J. Crutzen and E. F. Stoermer, "The 'Anthropocene'", *Global Change Newslett.*, 2000, **41**, 17–18.

2. S. J. Arrhenius, On the influence of carbonic acid in the air upon the temperature of the ground, *Philos. Mag.*, 1896, **41**, 237–76.

3. C. D. Keeling, The concentration and isotopic abundances of carbon dioxide in the atmosphere, *Tellus*, 1960, **12**, 200–203.

4. J. Hansen, D. Johnson, A. Lacis, S. Lebedeff, P. Lee, D. Rind and G. Russell, Climate impact of increasing atmospheric carbon dioxide, *Science*, 1981, **213**, 957–966, DOI: 10.1126/science.213.4511.957.

5. M. Meinshausen, N. Meinshausen, W. Hare, S. C. B. Raper, K. Frieler, R. Knutt, D. J. Frame and M. R. Allen, *Nature*, 2009, **458**, 1158–1163.

6. M. R Allen, D. J. Frame, C. Huntingford, C. D. Jones, J. A. Lowe, M. Meinshausen and N. Meinshausen, *Nature*, 2009, **458**, 1163–1166.

7. BP, *Statistical Review of World Energy*, 2013; www.bp.com/en/global/corporate/about-bp/statistical-review-of-world-energy-2013.html (accessed October 2013).

8. IEA International Energy Agency, *World Energy Outlook*, 2013; www.worldenergyoutlook.org (accessed October 2013).
9. C. Wilson, A. Grubler, K. S. Gallagher and G. F. Nemet, *Nature Climate Change*, 2012, **2**, 780–788.
10. D. McLaren, *Process Safety Environ. Protect*, 2012, **90**, 489–500.
11. International Panel on Climate Change, Carbon and other biogeochemical cycles, ch. 6, in *IPCC Fifth Assessment Report AR5*, in press; www.ipcc.ch.
12. D. Woolf, J. E. Amonette, F. Alayne Street-Perrott, J. Lehmann and S. Joseph, Sustainable biochar to mitigate global climate change, *Nature Commun.*, 2010, **1**(56), DOI: 10.1038/ncomms1053.
13. K. J. Anderson-Teixeira, S. C. Davis, M. D. Masters and E. H. DeLucia, Changes in soil organic carbon under biofuel crops, *GCB Bioenergy*, 2009, **1**(1), 75–96.
14. G. Berndes, M. Hoogwijk and R. van den Broek, The contribution of biomass in the future global energy supply: a review of 17 studies, *Biomass Bioenergy*, 2003, **25**(1), 1–28.
15. S. C. Davis, K. J. Anderson-Teixeira and E. H. DeLucia, Life-cycle analysis and the ecology of biofuels, *Trends Plant Sci.*, 2009, **14**(3), 140–146.
16. D. Weisser, A guide to life-cycle greenhouse gas (GHG) emissions from electric supply technologies, *Energy*, 2007, **32**(9), 1543–1559.
17. IEA *Bioenergy – a sustainable and reliable energy source*. 2009.
18. V. Dornburg, D. van Vuuren, G. van de Ven, H. Langeveld, M. Meeusen and M. Banse, Bioenergy revisited: key factors in global potentials of bioenergy, *Energy Environ. Sci.*, 2010, **3**(3), 258–267.
19. T. W. R. Powell and T. M. Lenton, Future carbon dioxide removal via biomass energy constrained by agricultural efficiency and dietary trends, *Energy Environ. Sci.*, 2012, **5**(8), 8116–8133.
20. M. Ashley, Nanomaterials and processes for carbon capture and conversion into useful by products, *Greenhouse Gases Sci. Technol.*, 2010, **2**(6), 419–444.
21. P. Styring, H. de Coninck, H. Reith and K. Armstrong, Carbon capture and utilisation in the green economy CO₂Chem, Sheffield, 2012, http://co2chem.co.uk/wp-content/uploads/2012/06/CCU in the green economy report.pdf.
22. P. Styring, Roadmap for the future of CCUS. CO₂Chem, Sheffield, 2012, http://co2chem.co.uk/co2chem-roadmap-spring-2012.
23. B. P. McGrail, H. T. Schaef, A. M. Ho, Y.-J. Chien, J. J. Dooley and C. L. Davidson, Potential for carbon dioxide sequestration in flood basalts, *J. Geophys. Res.*, 2006, **111**(B12201).
24. J. M. Matter and P. B. Kelemen, Permanent storage of carbon dioxide in geological reservoirs by mineral carbonation, *Nature Geosci.*, 2009, **2**(12), 837–841.
25. K. Z. House, D. P. Schrag, C. F. Harvey and K. S. Lackner, Permanent carbon dioxide storage in deep-sea sediments, *Proc. Natl. Acad. Sci. U. S. A.*, 2006, **103**, 12291–12295.

26. J. S. Levine, J. M. Matter, D. Goldberg, A. Cook and K. S. Lackner, Gravitational trapping of carbon dioxide in deep sea sediments: permeability, buoyancy, and geomechanical analysis, *Geophys. Res. Lett.*, 2007, **34**, L24703, DOI: 10.1029/2007GL031560.

27. IPCC, *IPCC Special Report on Carbon Dioxide Capture and Storage*, ed. B. Metz, Cambridge University Press, Cambrdge, K and New York, 2005, pp. 442.

28. J. Bradshaw, S. Bachu, D. Bonijoly, R. Burruss, N. P. Christensen, O. M. Mathiassen, *Discussion paper: reviewing and identifying standards with regards to CO$_2$ storage capacity measurement*, 2005, Carbon Sequestration Leadership Forum, http://www.cslforum.org/publications/documents/Oviedo/TaskForce_Reviewing_Ident_Stand_CO2_Storage_Cap.pdf.

29. National Energy Technology Lab (NETL), *The USA Carbon Utilization and Storage Atlas*, 4th edn, 2012; www.netl.doe.gov (accessed December 2012).

30. E. K. Halland, W. T. Johansen and F. Riis, *CO$_2$ Storage Atlas Norwegian North Sea*, 2012, Norwegian Petroleum Directorate.

31. Scottish Carbon Capture Storage (SCCS), *Opportunities for CO$_2$ Storage around Scotland*, 2009; www.sccs.org.uk (accessed January 2014).

32. CO$_2$ Stored, www.co2Stored.co.uk (accessed October 2013).

33. M. Jin, *Soc. Petrol. Eng. J.*, 2012, **17**(4), 1108–1118.

34. R. J. Stewart, V. Scott, R. S. Haszeldine, D. Ainger and S. Argent, The feasibility of a European wide integrated CO$_2$ transport network, *Int. J. Greenhouse Gas Control*, 2014, DOI: 10.1002/ghg.1410.

35. E. J. Stone, J. A. Lowe and K. P. Shine, The impact of carbon capture and storage on climate, *Energy Environ. Sci.*, 2009, **2**(1), 81–91.

36. G. Shaffer, Long-term effectiveness and consequences of carbon dioxide sequestration, *Nature Geosci.*, 2010, **3**(7), 464–467.

37. N. E. Vaughan and T. M. Lenton, A review of climate geoengineering proposals, *Climatic Change*, 2011, **109**, 745–790.

38. The Royal Society, *Geoengineering the Climate. Science, governance and Uncertainty*, The Royal Society, London, September 2009; royalsociety.org/uploadedFiles/Royal_Society_Content/policy/publications/2009/8693.pdf (accessed January 2014).

39. N. L. Harris, S. Brown, S. C. Hagen, S. S. Saatchi, S. Petrova, W. Salas, M. C. Hansen, P. V. Potapov and A. Lotsch, Baseline map of carbon emissions from deforestation in tropical regions, *Science*, 2012, **336**(6088), 1573–1576.

40. R. Jandl, M. Lindner, L. Vesterdal, B. Bauwens, R. Baritz, F. Hagedorn, D. W. Johnson, K. Minkkinen and K. A. Byrne, How strongly can forest management influence soil carbon sequestration?, *Geoderma*, 2007, **137**(3–4), 253–268.

41. F. Worrall, M. Bell and A. Bhogal, Assessing the probability of carbon and greenhouse gas benefit from the management of peat soils, *Sci. Total Environ.*, 2010, **408**(13), 2657–2666.

42. G. B. Bonan, Forests and climate change: forcings, feedbacks, and the climate benefits of forests, *Science*, 2008, **320**(5882), 1444–1449.
43. J. Grace, J. S. José, P. Meir, H. S. Miranda and R. A. Montes, Productivity and carbon fluxes of tropical savannas, *J. Biogeogr.*, 2006, **33**, 387–400.
44. S. Shackley, S. Sohi and R. Ibarrola, Biochar, tool for climate change mitigation and soil management, in *Geoengineering Responses to Climate Change*, ed. T. M. Lenton and N. E. Vaughan, Springer, 2013, 73–140.
45. S. Shackley and S. Sohi, *An Assessment of the Benefits and Issues associated with the Application of Biochar to Soil*, United Kingdom Department for Environment, Food and Rural Affairs, and Department of Energy and Climate Change, 2011.
46. S. Shackley, J. Hammond, J. Gaunt and R. Ibarrola, The feasibility and costs of biochar deployment in the UK, *Carbon Manage*, 2011, **2**(3), 335–356.
47. J. Hammond, S. Shackley, S. Sohi and P. Brownsort, Prospective life cycle carbon abatement for pyrolysis biochar systems in the UK, *Energy Policy*, 2011, **39**, 2646–2655.
48. S. Meyer, R. M. Bright, D. Fischer, H. Schulz and B. Glaser, Albedo impact on the suitability of biochar systems to mitigate global warming, *Environ. Sci. Technol.*, 2012, **46**(22), 12726–12734.
49. L. Genesio, F. Miglietta, E. Lugato, S. Baronti, M. Pieri and F. P. Vaccari, Surface albedo following biochar application in durum wheat, *Environ. Res. Lett.*, 2011, **7**, 014025.
50. International Energy Agency, *Technology Roadmap: Bioenergy for Heat and Power*, 2012.
51. International Energy Agency GHG R&D Programme, *Biomass CCS study*, 2009.
52. R. J. Finley, S. M. Frailey, H. E. Leetaru, O. Senel, M. L. Couëslan and M. Scott, *Energy Procedia*, 2013, **37**, 6149–6155.
53. P. Luckow, M. A. Wise, J. J. Dooley and S. H. Kim, Large-scale utilization of biomass energy and carbon dioxide capture and storage in the transport and electricity sectors under stringent CO_2 concentration limit scenarios, *Int. J. Greenhouse Gas Control*, 2010, **4**(5), 865–877.
54. J. Koornneef, A. Ramírez, W. Turkenburg and A. Faaij, The environmental impact and risk assessment of CO_2 capture, transport and storage – An evaluation of the knowledge base, *Int. J. Greenhouse Gas Control*, 2012, **11**, 117–132.
55. N. McGlashan, M. Workman, B. Caldecott and N. Shah, *Negative emissions technologies*, Grantham Institute Briefing paper 8. Imperial College, London, 2012.
56. D. Voll, A. Wauschkuhn, R. Hartel, M. Genoese and W. Fichtner, Cost estimation of fossil power plants with carbon dioxide capture and storage, *Energy Procedia*, 2012, **23**, 333–342.
57. R. Lal, M. Griffin, J. Apt, L. Lave and M. G. Morgan, Managing soil carbon, *Science*, 2004, **304**(5669), 393.

58. S. E. Strand and G. Benford, Ocean sequestration of crop residue carbon: recycling fossil fuel carbon back to deep sediments, *Environ. Sci. Technol.*, 2009, **43**(4), 1000–1007.

59. A. G. Glover, H. Wiklund, S. Taboada, C. Avila, J. Cristobo, C. R. Smith, K. M. Kemp, A. J. Jamieson and T. G. Dahlgren, *Proc. R. Soc. London, Ser. B*, 2013, 280, DOI: 10.1098/rspb.2013,1390.

60. J. C. M. Pires, M. C. M. Alvim-Ferraz, F. G. Martins and M. Simões, Carbon dioxide capture from flue gases using microalgae: engineering aspects and biorefinery concept, *Renewable Sustain. Energy Rev.*, 2012, **16**(5), 3043–3053.

61. M. S. A. Rahamana, L. H. Chengb, X. H. Xub, L. Zhanga and H. L. Chena, A review of carbon dioxide capture and utilization by membrane integrated microalgal cultivation processes, *Renewable Sustain. Energy Rev.*, 2011, **15**(8), 4002–4012.

62. C. K. Savile and J. J. Lalonde, Biotechnology for the acceleration of carbon dioxide capture and sequestration, *Curr. Opin. Biotechnol.*, 2011, **22**(6), 818–823.

63. L. Gustavsson, K. Pingoud and R. Sathre, Carbon dioxide balance of wood substitution: comparing concrete-and wood-framed buildings, Mitigation Adaptation Strateg, *Global Change*, 2006, **11**(3), 667–691.

64. Virgin, *Earth Challenge*, 2011; www.virgin.com/unite/entrepreneurship/virgin-earth-challenge-announces-leading-organisations (accessed October 2013).

65. S. Brandani, Carbon dioxide capture from air: a simple analysis, *Energy Environ.*, 2012, **23**(2), 319–328, DOI: 10.1260/0958-305X.23.2-3.319.

66. Global Thermostat, 2013; globalthermostat.com (accessed October 2013).

67. K. S. Lackner, Capture of carbon dioxide from ambient air, *Eur. Phys. J. – Special topics*, 2009, **176**(1), 93–106.

68. K. S. Lackner, S. Brennan, J. M. Matter, A.-H. Alissa Park, A. Wright and B. van der Zwaan, *Proc. Natl. Acad. Sci. U. S. A.*, 2012, **109**(33), 13156–13162.

69. Kilimanjaro Energy; www.kilimanjaroenergy.com (accessed October 2013).

70. Carbon Engineering; carbonengineering.com (accessed October 2013).

71. D. W. Keith, Why capture CO_2 from the atmosphere?, *Science*, 2009, **325**(5948), 1654–165.

72. R. M. Socolow, M. Desmond, R. Aines, J. Blackstock, O. Bolland, T. Kaarsberg, N. Lewis, M. Mazzotti, A. Pfeffer, K. Sawyer, J. Siirola, B. Smit, J. Wilcox, *Direct air capture of CO_2 with chemicals*. A Technology Assessment for the APS Panel on Public Affairs, 2011, American Physical Society.

73. K. Z. House, A. C. Baclig, M. Ranjan, E. A. van Nierop, J. Wilcox and H. J. Herzog, Economic and energetic analysis of capturing CO_2 from ambient air, *Proc. Natl. Acad. Sci. U. S. A.*, 2011, **108**(51), 20428–20433.

74. O. Boucher and G. A. Folberth, New directions: atmospheric methane removal as a way to mitigate climate change?, *Atmos. Environ*, 2010, **44**, 3343–3345.
75. R. Schuiling and P. Krijgsman, Enhanced weathering: an effective and cheap tool to sequester CO_2, *Climatic Change*, 2006, **74**(1), 349–354.
76. P. Köhler, J. Hartmann and D. A. Wolf-Gladrow, Geoengineering potential of artificially enhanced silicate weathering of olivine, *Proc. Natl. Acad. Sci. U. S. A.*, 2010, **107**(47), 20228–20233.
77. Smithsonian 2011, http://www.smithsonianmag.com/science-nature/building-a-better-world-with-green-cement-81138/?all, Royal Society 2011, http://royalsociety.org/news/Royal-Society-invests-Novacem/.
78. International Energy Agency (IEA), *Cement Technology Roadmap*, 2009; www.iea.org (accessed October 2013).
79. NETL, *CO2-EOR Primer, Carbon Dioxide Enhanced Oil Recovery*, 2010; www.netl.doe.gov/technologies/oil-gas/publications/EP/CO2_EOR_Primer.pdf (accessed October 2013).
80. P. Jaramillo, V. M. Griffin and S. T. McCoy, Life cycle inventory of CO_2 in an enhanced oil recovery system, *Environ. Sci.Technol.*, 2009, **43**, 8027–8032.
81. J. Stewart and R. S. Haszeldine, Life cycle analysis for CO2-EOR, offshore North Sea, *Environ. Sci. and Technol.*, submitted.
82. R. S. Haszeldine, Carbon capture and storage, how green can black be?, *Science*, 2009, **325**, 1647–1652.
83. V. Scott, S. Gilfillan, N. Markusson, H. Chalmers and R. S. Haszeldine, Last chance for carbon capture and storage, *Nature Climate Change*, 2012, **3**, 105–111.
84. Jevons effect; en.wikipedia.org/wiki/Jevons_paradox (accessed October 2013).

The Global Potential for Carbon Dioxide Removal

TIMOTHY M. LENTON

ABSTRACT

The global physical potential of different methods of carbon dioxide removal (CDR) from the atmosphere is reviewed. A new categorisation into plant-based, algal-based and alkalinity-based approaches to CDR is proposed. Within these categories, the key flux-limiting resources for CDR are identified and the potential CO_2 removal flux that each technology could generate is quantitatively assessed – with a focus on the present, 2050 and 2100. This reveals, for example, that use of waste nutrient flows to feed macro-algae for biomass energy with carbon capture and storage (algal BECCS), shows significant CDR potential, without needing the large land areas or freshwater supplies of plant biomass energy crops. Adding up the potentials of different CDR methods, the total CDR potential at present is 1.5–3 PgC yr^{-1} (Petagram of carbon per year), comparable in size to either the natural land or ocean carbon sinks. Already 0.55–0.76 PgC yr^{-1} of this potential has been realised through afforestation and inadvertent ocean fertilisation. The total CDR potential (without including direct air capture) grows such that by mid-century it is 4–9 PgC yr^{-1} and by the end of the century it is 9–26 PgC yr^{-1}, comparable with current total CO_2 emissions of 10 PgC yr^{-1}. The CDR that can be realised under social, economic and engineering constraints is always going to be less than the physical potential. Nevertheless, if combined with reducing CO_2 emissions (conventional mitigation), CDR has the physical potential to help stabilise atmospheric CO_2 by the middle of this century.

Issues in Environmental Science and Technology, 38
Geoengineering of the Climate System
Edited by R.E. Hester and R.M. Harrison
© The Royal Society of Chemistry 2014
Published by the Royal Society of Chemistry, www.rsc.org

1 Introduction

The global carbon cycle is currently out of balance. Human fossil fuel burning and land use change activities are producing a combined source of CO_2 to the atmosphere of around ~ 10 PgC yr^{-1} (petagrams or Pg $= 10^{15}$).[1] This is causing atmospheric CO_2 concentration to rise at ~ 2 ppm yr^{-1} and causing carbon to accumulate in the ocean and in land ecosystems.[1] The rise in atmospheric CO_2 concentration is in turn making the single largest contribution to increasing global temperatures. Thus, in order to minimise the rise of global temperature, the rise of atmospheric CO_2 must be halted (or sunlight reflection methods of geoengineering must be deployed).

In simple terms, stabilizing atmospheric CO_2 concentration demands that carbon sinks (removal fluxes from the atmosphere) match carbon sources (fluxes to the atmosphere). Lowering atmospheric CO_2 concentration demands that sinks exceed sources. The conventional policy framework for achieving stabilization is to reduce CO_2 emissions to match natural (land and ocean) sinks, and then to reduce CO_2 emissions to zero (at least as fast as natural sinks decay). The cumulative carbon emission will then determine the resulting change in global temperature, called the "cumulative warming commitment".[2] This policy approach poses a profound collective challenge to transform the current exponential increase in CO_2 emissions ($\sim 2\%$ yr^{-1} over the past 25 years and $> 3\%$ yr^{-1} at the beginning of the 21st century)[3,1] into a comparable or greater rate of decrease in CO_2 emissions. The required transition of the global energy system must start soon and be completed within decades, if global warming is to be restricted to less than 2 °C above pre-industrial.[2,4,5] Already it demands rates of technological and economic change that may be politically unachievable.[3]

Hence there is growing interest in the potential for deliberate carbon dioxide removal from the atmosphere to augment reductions in CO_2 emissions. Carbon dioxide removal (CDR) – or Negative Emissions Technologies (NETs) – describes a suite of methods that remove CO_2 from the ambient air by biological, chemical or physical means and store the resulting carbon in long lived reservoirs. If both anthropogenic emissions of CO_2 are reduced and CO_2 sinks are created, then the rise of atmospheric CO_2 concentration (and global temperature) can be halted sooner and at a lower level than by reducing emissions alone. Indeed, CDR is already implicit in most scenarios to stay under 2 °C of global warming above pre-industrial, including the IPCC RCP2.6 scenario, where it takes the form of widespread biomass energy with carbon capture and storage (BECCS).[6] CDR effectively reduces the cumulative carbon emission and hence reduces the corresponding global warming commitment.[2] Ultimately, CDR could be used to bring atmospheric CO_2 concentration down to whatever is considered a safe level. CDR may also be used to counter-balance some "essential" or "unavoidable" fossil fuel CO_2 emissions, without increasing the CO_2 concentration.

However, most CDR technologies are more expensive than most conventional emissions reduction options, and hence are unlikely to be used until after the cheaper mitigation options. Furthermore, because CDR offers the option to reduce atmospheric CO_2 concentrations at some later time, this may reduce the urgency to start cutting emissions now. Indeed, knowing that global temperature change lags changes in radiative forcing, some studies have framed CDR as allowing a temporary overshoot in atmospheric CO_2 concentrations above safe levels,[7] without overshooting a corresponding safe temperature target. This would of course be a risky strategy if the imagined future potential for CDR cannot be realised in practice.

There are thus several reasons to want to know the global potential for CDR from a scientific perspective, before also thinking about the engineering, the costs and the social acceptability of CDR technologies. The most critical factor in determining the global potential of CDR is the flux of CO_2 removal that can be achieved at a given time. The achievable CDR flux, together with the anthropogenic emissions flux and natural sinks fluxes, determines whether CO_2 concentration can be stabilised, reduced, or will continue rising, at a given time. In the longer term, the total storage capacity for removed CO_2, together with the total cumulative CO_2 emission, will determine how much anthropogenic CO_2 remains in the atmosphere–ocean system, and therefore the long-term concentration of CO_2 and the corresponding warming.[2,8–10] Also important in the long term is whether there is leakage of CO_2 from the storage reservoirs back to the atmosphere, and if so, at what rate.

The various methods available for carbon dioxide removal (see Figure 1) have been summarised in previous work.[11,12] They can be categorised into biological, chemical and physical approaches, or land and ocean based approaches. Here I suggest a new categorization into plant-based, algal-based, and alkalinity-based CDR approaches. The various CDR options include permanent afforestation and reforestation, biomass burial, biochar production, biomass energy with carbon capture and storage (BECCS), either from plant or algal material, ocean fertilization with macronutrients (nitrogen, phosphorus), or micronutrients (iron), enhanced weathering, ocean liming, and direct air capture (DAC). Current assessments suggest that land-based methods of CO_2 removal either *via* biological (photosynthesis) or chemical and physical means, have greater potential than ocean-based methods.[11,12] Furthermore, existing economic assessment suggests that land-based biological CDR has a better cost–benefit ratio than direct air capture of CO_2 using chemical and physical means (although direct air capture would take up far less land space).[13]

Before getting into the specifics of the different CDR pathways, let us note some general overarching constraints for the generation of any CDR flux. All CDR fluxes can be viewed as depending upon: (i) a supply of some limiting resource(s) to capture CO_2; (ii) a yield of carbon per unit input of limiting resource; and (iii) a conversion efficiency of that carbon to long-lived storage, including a supply of resource(s) to achieve that capture. For example, biomass energy with carbon capture and storage (BECCS) using terrestrial plants depends upon a supply of land area, along with nutrients and freshwater, a yield of carbon per unit land area (which will depend on

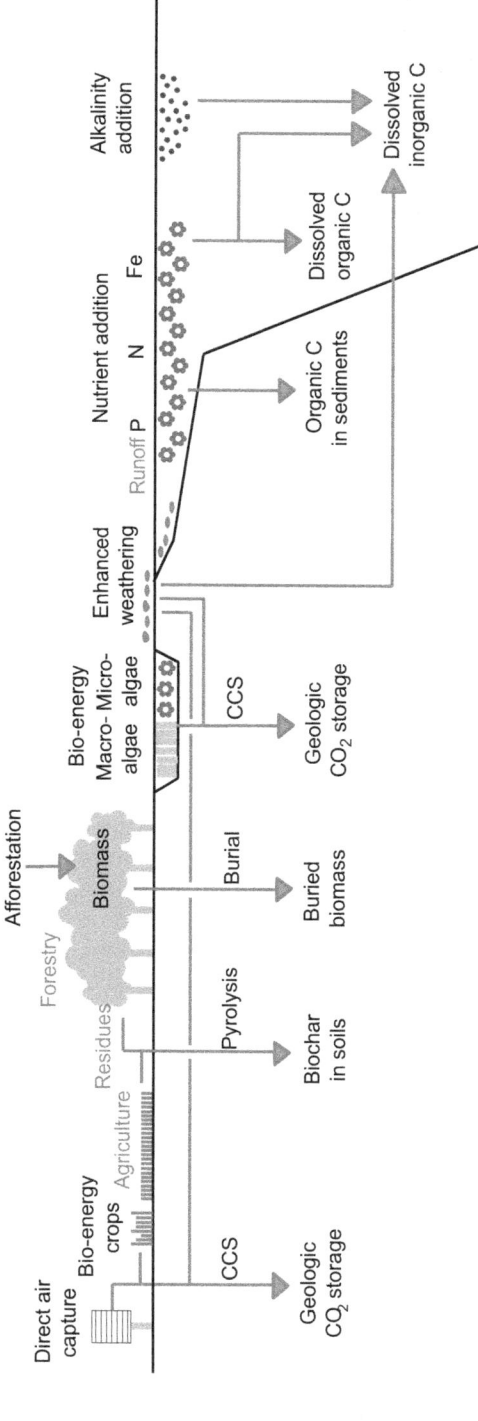

Figure 1 Methods of carbon dioxide removal (CDR).

nutrient and water availability and the chosen biomass crop), and a conversion efficiency to liquid CO_2 stored underground, which takes into account the energy penalty for the conversion and storage process. For each CDR proposal, it may be possible to identify a rate-limiting step in this sequence that limits the CDR flux that can be generated.

The aim in this chapter is to provide an up-to-date review of the global, physical CDR flux potential of leading candidate technologies up to the century timescale. I do not dwell on CDR proposals where existing assessments indicate they have minimal potential,[11] such as enhancing ocean upwelling or down-welling. I assess the flux potential of plant-based CDR, algal-based CDR and alkalinity-based CDR and then the combined total CDR potential, seeking to avoid double counting. Having quantified the physical potential for CDR, I then briefly discuss the implications for climate policy, the caveats, and some directions for future research.

2   Plant-based CDR

"Plant-based CDR" refers to all pathways where CO_2 is removed from the atmosphere by terrestrial plants and some of the resulting biomass flows are converted to stored carbon (see Figure 1). The simplest plant-based CDR pathway is to accumulate carbon in woody biomass through permanent afforestation, perhaps augmenting the sink by harvesting some of the biomass as wood products and thus maintaining the corresponding forestry plantations in a high growth phase.[14,15] Alternative suggestions are to deliberately bury wood in soils,[16] or crop residues in ocean sediments,[17,18] to store carbon. None of these pathways make use of the chemical energy in biomass. Alternatively, if energy is extracted from biomass, some of the associated carbon can in principle be captured and stored as biochar (from pyrolysis of biomass),[19] or as CO_2 (from fermentation, gasification or combustion processes),[20,21] either as liquid CO_2 in geological storage or in seawater when combined with a balanced source of alkalinity. The feedstocks for these bioenergy CDR pathways could include deliberately grown energy crops, forestry wood that is surplus to other uses, and residues (*i.e.* waste products) from agriculture, energy crops and forestry. The different end storage forms for carbon can be derived from the same land and/or the same biomass feedstock, so in estimating global CDR flux potential one must be careful to avoid double counting. Hence I first consider potential supplies of "new" land and the afforestation or bioenergy crops that could be grown on it. Then I consider additional biomass flows from forestry and agriculture.

2.1   Resource Supplies

For plant-based CDR, a key limiting resource supply is land area, especially in a world where demand for food is increasing rapidly and there is a desire to preserve natural ecosystems, including forests. Abandoned agricultural land is a prime target for plant-based CDR as it is generally the most productive land. It could either be afforested or devoted to biomass energy crops. Historically, since the early 1960s there has been little net change in land area under cultivation despite a doubling of population,[22] but some agricultural land has been abandoned, whilst new land went under cultivation elsewhere. Looking ahead, most scenarios project a continuing supply of abandoned cropland,[23] of 0.6–1.3 Gha in 2050 and about double this in 2100.[24] The corresponding supply rate of abandoned agricultural land fluctuates over 0–17 Mha yr^{-1} across the scenarios.[25] However, if gains in agricultural efficiency of around 1% per year cannot be sustained in the face of growing food demand, global land area under cultivation will have to increase – leaving little high-productivity land for plant-based CDR.[26] Supplies of nutrients and freshwater could also limit plant-based CDR potential – especially when trying to achieve high productivity and yields with dedicated biomass energy crops.[27]

Low-productivity land (including grazed grassland) is projected to dwindle in area, and to have negligible potential for deliberate biomass growth.[24] However, one ambitious proposal suggests that large areas of very low-productivity desert in Australia and the Sahara could be irrigated by desalination of seawater, and forests grown there.[28] In that case, the key limiting resources will be the supply of energy to desalinate seawater and pipe it, and the corresponding nutrients needed to establish forest soils.

2.2 Afforestation and Reforestation

The conversion of unforested land to permanent forest creates a net carbon sink and a store of carbon in the biomass of the trees and in the soil, although there can be transient (and even net) loss of carbon from soil depending on location. The afforestation CDR flux grows both as planted trees approach their peak rates of carbon accumulation and as progressively more land is subject to planting. Once a forest reaches maturity, the sink declines to zero with respiratory carbon losses matching photosynthetic carbon uptake, although recent studies point to a persistent carbon sink in old growth forests.[29] By harvesting carbon in the form of wood products and replanting, forestry plantations can be maintained in a higher average yield state, thus increasing the CDR flux.[15] Yields of carbon for permanent afforestation are of the order 1 MgC ha^{-1} yr^{-1},[15] with average values of 0.8–1.6 MgC ha^{-1} yr^{-1} used in global projections.[14,15]

Large afforestation programs have already been undertaken, with an estimated 264 Mha afforested in 2010.[30] In China alone, the corresponding CDR flux is estimated to have been 0.19 PgC yr^{-1} over 1988–2001.[31] If the 264 Mha of existing plantations are accumulating carbon at an average rate of 0.8–1.6 MgC ha^{-1} yr^{-1},[14,15] then the corresponding CDR is already 0.21–0.42 PgC yr^{-1}. Conceivably this is an under-estimate as yield can be considerably greater in the tropics.

Afforestation area increased at $\sim$5 Mha yr^{-1} over 2005–2010,[30] which simply extrapolating forward could produce an additional 200 Mha afforested by 2050. On the resulting total area of 464 Mha, the afforestation CDR flux in 2050 could be 0.37–0.74 PgC yr^{-1} (assuming 0.8–1.6 MgC ha^{-1} yr^{-1}). This is in the range of projections for 2050[32] with more detailed models of 0.2–1.5 PgC yr^{-1},[15,25,33,34] which is constrained by the supply of abandoned cropland.

In 2100, afforestation CDR projections from detailed models range over 0.3–3.3 PgC yr^{-1},[15,25] again constrained by the supply of abandoned cropland. Much larger technical potentials have been suggested on shorter time horizons (*e.g.* ∼10 PgC yr^{-1} in 2035)[35,36] but these imply major conflicts with food production and/or the preservation of natural ecosystems and hence are ignored here.[32]

2.3 Bioenergy Crop Supplies

Woody bioenergy crops (*e.g.* Pinus, Eucalyptus – globally the two main species), have higher achieved yields than afforestation of ∼1.5–7 MgC ha^{-1} yr^{-1} (assuming ∼0.5 gC g^{-1} average carbon content of wood).[37,38] Even higher global average yield levels are assumed in some projections, ranging over 1.5–15 MgC ha^{-1} yr^{-1} or 8–10.5 MgC ha^{-1} yr^{-1},[37,39] which seem ambitiously high.[14] Other energy crops generally have yields less than or equal to woody crops.

Current bioenergy supplies meet about 6% of global energy demand but only a small fraction of this is in the form of modern biomass energy crops, amenable to large-scale capture and storage technology.

In 2050, it has been estimated that bioenergy crops could supply 40–330 EJ yr^{-1} (exajoules or EJ = 10^{18}),[37,40] which assumes that the plantation area will range over 390–750 Mha, with yields typically 4–7.5 MgC ha^{-1} yr^{-1} and global production of around 2.5–3 PgC yr^{-1} (assuming ∼0.5 gC g^{-1}).[37] Alternative estimates are that if 1% yr^{-1} gains in agricultural efficiency can be sustained to 2050, with the expected trend toward higher meat diets, 1.7 PgC yr^{-1} of energy crop could be produced in 2050.[26] If in addition there were a reversion to lower meat diets, bioenergy crop supplies could increase to 3.4 PgC yr^{-1} in 2050.[26] However, this would demand massive nutrient inputs. For each ∼1 PgC yr^{-1}, roughly 20% of present global fertiliser nitrogen would be required, based on switchgrass as the crop.[27] Expressed another way this is a ratio of nitrogen input to carbon output of C : N ∼58, which with typical grass N : P ∼15–20, corresponds to C : P ∼1000, and is consistent with around 20% of global fertiliser phosphate inputs (of 0.39–0.45×10^{12} mol P yr^{-1}) also being required for each ∼1 PgC yr^{-1}.

By 2100, integrated assessments of bioenergy potential constrained by the supply of suitable land area tend to be roughly double what they are in 2050.[24,37] The greatest potential is on abandoned agricultural land, with one study giving a range of 240–850 EJ yr^{-1} in 2100 for woody energy crops,[24] corresponding to 6–21 PgC yr^{-1} (assuming 20 GJ Mg^{-1} and 0.5 gC g^{-1}). However, the required nitrogen and phosphorus inputs to sustain this would be roughly 100–400% of present fertiliser production, unless efficient nutrient recycling systems can be implemented.

2.4 Additional Biomass Supplies

With large areas of the land surface already managed, wood supplies from forests and waste biomass flows from agriculture and forestry can provide major sources of carbon for conversion to long-lived forms.

Based on year 2000 figures,[41] global wood removals from forests were ~ 1 PgC yr^{-1}, roughly half of which went to wood fuel and half to industrial uses, but it is unclear whether any of the ~ 0.5 PgC yr^{-1} used as wood fuel is available for capture and storage. Global flows of carbon in unused residues from cropland were ~ 0.6 PgC yr^{-1} (assuming 0.4 gC g^{-1}).[41] Alternative estimates of maximum current potential agricultural residues are 0.28 PgC yr^{-1} from rice, 0.18 PgC yr^{-1} from other cereals, 0.13 PgC yr^{-1} from sugar cane waste bagasse and field trash, and 0.19 PgC yr^{-1} from manures, giving 0.78 PgC yr^{-1} in total.[42] In addition, year 2000 felling losses from forestry were ~ 0.33 PgC yr^{-1} (assuming 0.5 gC g^{-1}).[41] Thus combined total residues are in the range 0.9–1.1 PgC yr^{-1} which, with the addition of all wood fuel, gives ~ 1.5 PgC yr^{-1}.

In 2050, forecast bioenergy supply from surplus forest biomass is 60–100 EJ yr^{-1},[40] which corresponds to 1.5–2.5 PgC yr^{-1} (assuming 20 GJ Mg^{-1} and 0.5 gC g^{-1}), and would represent a significant increase over current global wood removals of ~ 1 PgC yr^{-1}.[41] In 2050, the supply of agricultural and forest residues is estimated at 30–180 (mean 100) EJ yr^{-1},[40] which corresponds to 0.8–4.8 (mean 2.7) PgC yr^{-1} (assuming 15 GJ Mg^{-1} and 0.4 gC g^{-1} for residues). The lower end of this range is less than unused crop residues and felling losses at present,[41] but the upper estimate looks unrealistic. Alternative estimates for 2050, including manure and food waste as potential feedstocks, give total biomass "waste" streams of 2.2–2.7 PgC yr^{-1} (across four scenarios) as a more reasonable upper limit.[26]

2.5 Conversion Routes and Efficiencies

Having established the supply of carbon in biomass the next key question is; how much can be converted to long-term storage?

If one leaves biomass in permanent forests and their soils (where previously the land stored less carbon) the conversion efficiency is often treated as 100%, although natural disturbances such as pests and fire that reduce carbon storage cannot be completely prevented.[43] Biomass burial has a conversion efficiency >97% based on data for solid wood in landfill sites, where <3% of the carbon is converted to CH_4 and CO_2 in a roughly 1 : 1 ratio.[44] Burial of biomass in the deep ocean is also assumed to be near 100% efficient.

BECCS technologies generally lend themselves to relatively uniform feedstock such as dedicated bioenergy crops. There are several pathways for converting biomass carbon to captured CO_2 including: (i) biomass combustion with flue gas CO_2 capture (which can also use mixed feedstock); (ii) biomass gasification then CO_2 capture (with an optional CO shift) before combustion or conversion to fuel; (iii) air separation of pure O_2 for biomass combustion with CO_2 capture; (iv) biomass fermentation to biofuel (sometimes preceded by saccharification) with CO_2 capture; and (v) biomass conversion to biofuel *via* the Fischer–Tropsch process with CO_2 capture.[21]

CO_2 capture potential varies considerably across these technologies (as do the offsets of fossil fuel burning). Carbon capture yields of 90% (and possibly higher), with a corresponding $\sim$30% energy yield as electricity, are claimed for (ii) a biomass integrated gasification combined cycle (BIGCC) with CCS.[45,46] However, other authors estimate only a 55% carbon yield with 25% energy yield as electricity for the same type of system.[47] Higher energy yields are estimated in the form of hydrogen production (55%) or heat production (80%).[45] Lowest carbon capture yields are for converting sugar cane to ethanol (iv), which leaves 67% of the carbon in the ethanol and releases 33% as CO_2. However, future biofuel is likely to be dominated by lignocellulosic crops, for which processing by saccharification and fermentation (iv) or Fischer–Tropsch (v) leaves around half of the carbon in the fuel, with carbon yields of (iv) $\sim$13% or (v) $\sim$41% as high purity CO_2.[48]

Biochar production lends itself to biomass residues from agriculture and forestry and other mixed feedstocks like food waste and manure, providing a convenient recycling mechanism for organic wastes. Charcoal is typically produced by pyrolysis of biomass although thermo-catalytic depolymerisation has also been demonstrated. The carbon and energy yields of biochar production vary greatly with the temperature of pyrolysis. In systems optimised for biochar yield,[49] up to 63% carbon capture is possible *via* pressurised flash pyrolysis, with an energy yield of around 35% in gas (59% of the energy is left in the char and 6% lost). A more conservative figure is $\sim$50% carbon capture with a similar energy yield.[42] When returned to soil as biochar a significant (but debated) fraction of the carbon in charcoal, *e.g.* 85%, is long-term resistant to biological decay.[42] Thus, the maximum overall conversion efficiency for biochar is probably around 50%.

2.6 Combined CDR Potential

The combined potential for plant-CDR (see Table 1) depends on the choice of land use and the choice of conversion pathways for biomass supplies. To maximise CDR potential, abandoned agricultural land should be devoted to dedicated bioenergy crops rather than afforestation. Then burial of all biomass could probably maximise CDR, but runs the risk of damaging ecosystems by withdrawing carbon and nutrients and, with no energy benefit, is unlikely to be favoured. Instead, uniform feedstock including dedicated bioenergy crops and surplus wood from forestry could be directed to BIGCC with CCS as it offers the potential to fix up to 90% of the carbon from the feedstock, whilst also yielding energy. For heterogeneous feedstock, including various biomass "waste" streams, pyrolysis and return of biochar to the soil has the potential to sequester around 50% of the carbon content of the feedstock, whilst also returning some associated nutrients to the soil.

The maximum combined plant-CDR potential at present is around 0.75–1.5 PgC yr^{-1} (see Table 1). Current afforestation CDR is conservatively 0.21–0.42 PgC yr^{-1}. The present potential for bioenergy CDR from

Table 1 Estimates of global plant-CDR flux potential by pathway for present, 2050 and 2100.

Carbon source	Carbon store	CDR flux (PgC yr^{-1})	Key conditions / assumptions	Reference
Present				
Afforestation	Biomass	0.21–0.42	264 Mha×0.8–1.6 MgC ha^{-1} yr^{-1}	30,32
Energy crops	BECCS	0.19–0.23	Sugarcane, ethanol and pulp mills	21
	Biochar	0.18	All biomass energy by pyrolysis	19
Forestry residues	Burial	0.33	0.65 Pg yr^{-1} of felling losses	32,41
	Biochar	0.16	All felling losses from forestry	32,41
Crop residues	Burial	0.18–0.6	1.5–5 Pg yr^{-1}, 30% removed	18,32,41
	Biochar	0.18	50% of unused crop residues	50
All residues	Biochar	0.16–0.34	original-revised estimates	19,32,41
Shifting cultivation	Biochar	0.21–0.35	All shifting cultivation fires	19,32,51
Charcoal	Biochar	0.01	All waste from charcoal making	19
All	Various	**0.75–1.5**	Total	
2050				
Afforestation	Biomass	0.2–1.5	All abandoned cropland	15,25,33,34
	Biomass	0.2–0.4	No expansion of afforested area	30,32
Energy crops	BECCS	1.25–1.5	390–750 Mha, 8–15 Mg ha^{-1} yr^{-1}	32,37
	BECCS	1.5–3.0	1.7–3.4 PgC yr^{-1} across 4 scenarios	26
Surplus wood	Biochar	0.75–1.25	60–100 EJ yr^{-1}, 20 GJ Mg^{-1}, 0.5 gC g^{-1}	32,40
All residues	Biochar	1.35	~100 EJ yr^{-1}, 15 GJ Mg^{-1}, 0.4 gC g^{-1}	32,40
	Biochar	1.1–1.35	2.2–2.7 PgC yr^{-1} across 4 scenarios	26
All	Various	**2.3–5.75**	Total (avoiding double counting)	
2100				
Afforestation	Biomass	0.3–3.3	SRES A2, B1, B2, A1b range	15,25
Energy crops	BECCS	5.4–19	240–850 EJ yr^{-1}, 20 GJ Mg^{-1}, 0.5 gC g^{-1}	24,32
	Biochar	5.5–9.5	180–310 EJ yr^{-1} all pyrolysed	19
Surplus wood	Biochar	0.75–1.25	2050 estimates (as above)	32,40
All residues	Biochar	1.1–1.35	2050 estimates (as above)	26
All	Various	**~5–20**	Total (avoiding double counting)	

sugarcane-based ethanol production and chemical pulp mills has been estimated at 0.19–0.23 PgC yr^{-1}.[21] Alternatively, if all "modern" biomass energy were converted to biochar by pyrolysis, the estimated CDR flux is 0.18 PgC yr^{-1}.[19] In addition, if global flows of crop residues from agriculture were all converted to biochar,[41] 0.3 PgC yr^{-1} could be removed,[50] or if 30% of these residues were buried in the deep ocean, 0.18 PgC yr^{-1} could be removed. In addition, the ~0.33 PgC yr^{-1} of felling losses from forestry,[41] could produce a CDR flux of 0.32 PgC yr^{-1} if all buried in soils or 0.16 PgC yr^{-1} as biochar. For the upper estimate, an additional 0.21–0.35 PgC yr^{-1} as biochar is included assuming "slash-and-char" shifting cultivation was adopted in place of current slash-and-burn practices.[19,32,51] Finally, there is 0.01 PgC yr^{-1} from wastes of charcoal production.[19] However, it is assumed that none of the wood currently being used as fuel is available for alternative conversion pathways (*e.g.* pyrolysis to biochar).

The maximum combined plant-CDR potential in 2050 could range over 2.3–5.75 PgC yr^{-1} (see Table 1). The lower estimate comes from a scenario in which gains in agricultural efficiency cannot be sustained and agricultural land has to expand in area, with current bioenergy crop land having to be reclaimed for food production. In that case, residues are maximised and their conversion to biochar can produce 1.35 PgC yr^{-1}. Present afforested area (with no further expansion) creates a minimum additional CDR flux of ∼0.2 PgC yr^{-1}, and wood removal from all managed forests could produce an additional 0.75 PgC yr^{-1} if converted to biochar. The upper estimate comes from a world in which agricultural efficiency continues to increase, the global trend towards eating more meat is reversed, and all abandoned agricultural land is devoted to bioenergy crops, which support a CDR flux of 3.0 PgC yr^{-1}. Wood removal is maximised giving 1.25 PgC yr^{-1} as biochar. Residues are somewhat less (2.2 PgC yr^{-1}) due to increases in agricultural efficiency, but produce a biochar CDR flux of 1.1 PgC yr^{-1}. Even though afforestation has not expanded, the current area supports an additional CDR flux ∼0.4 PgC yr^{-1}. Integrated assessment modelling of realised CDR potential in 2050 gives either 0–2.7 PgC yr^{-1} by BECCS or 0–1.1 PgC yr^{-1} by afforestation,[52] which, as would be expected, is less than the maximum potential estimates here. Furthermore, the freshwater and nutrient demands of removing several PgC yr^{-1} by BECCS and/ or afforestation may be "unrealistically high".[27]

The maximum combined plant-CDR potential in 2100 is particularly uncertain but could range over *circa* 5–20 PgCyr^{-1} (see Table 1). The lower limit comes from a maximum afforestation scenario (3.3 PgC yr^{-1}) with the addition of 2050 values for wood removal (0.75 PgC yr^{-1}) and forest and crop residue (1.1 PgC yr^{-1}) conversion to biochar. If, instead, one takes a lower limit for woody bioenergy crop production and converts it by BECCS (5.4 PgC yr^{-1}), plus the wood removal and residue biochar production fluxes, the total CDR flux is 7.25 PgC yr^{-1}. For an upper limit, the maximum woody bioenergy crop production with conversion by BECCS could yield up to ∼19 PgC yr^{-1} to which can be added wood removal and residues generated fluxes. The overall range spans earlier estimates of 5.5–9.5 PgC yr^{-1} potential biochar CDR flux (from 180–310 EJ yr^{-1} biomass energy supply in 2100).[19,37] Of course the realised potential would be expected to be lower, with integrated assessment modelling giving a BECCS potential of 0–5.4 PgC yr^{-1} in 2100.[52]

3   Algal-based CDR

"Algal-based CDR" refers here to all pathways in which CO_2 is removed from the atmosphere by algae (including cyanobacteria and macro-algae) and some of the resulting biomass flows are converted to stored carbon (see Figure 1). This categorisation groups ocean fertilization methods and algal bioenergy production with carbon capture and storage. The latter could be conducted in freshwater or salt water.

3.1 Resource Supplies

For algal-based CDR the input of nutrient(s) (and their subsequent recycling) is the key limiting resource, especially for cyanobacteria and micro-algae which have relatively low carbon-to-nutrient ratios compared with terrestrial plants. If algal-based CDR is attempted on land, the supply of suitable area for water containers also needs to be considered, whereas for ocean fertilization the supply of ships is important.

3.2 Algal BECCS

Micro-algae are an attractive source of bioenergy because their photosynthetic efficiency (10–20%) far exceeds that of land plants (1–2%) and they can achieve very high growth rates, doubling their biomass within a few hours under ideal conditions. Also many taxa grow in seawater, eliminating the need for freshwater inputs. Countering this is the problem that the typical carbon-to-nutrient ratio of micro-algae is much lower than that of land plants – especially woody plants – therefore the total nutrient demand to fix a given flux of carbon is much higher. Macro-algae have the advantage of higher carbon-to-nutrient stoichiometry than micro-algae and also have high productivity. Micro-algae can be cultivated in flow-through open "raceway" ponds or in closed "photobioreactors", whereas macro-algae can be cultivated in tidal-flat farms, nearshore farms, rope system farms, open-ocean farms or as floating seaweed.[53] However, it can take up to 20% of the energy captured to harvest algal biomass (greater than for terrestrial plants). To maximise photosynthetic carbon fixation in seawater, CO_2 needs to be continually resupplied by replenishing algae with CO_2-rich water, especially when grown in closed systems. Hence linking micro-algal biofuel production to high CO_2 concentration flue gases from power stations has been considered – but this only amounts to a CDR technology if the CO_2 is derived from biomass combustion, in which case algae can just be viewed as part of the capture technology. Instead, true algal-BECCS should be based on algae in water exchanging CO_2 with the free air. The subsequent conversion of biomass can involve biofuel production with CCS, anaerobic digesters producing methane (as well as CO_2) with subsequent combustion and CCS, and/or biomass gasification (BIGCC) with CCS.

Micro-algae in raceway ponds can capture up to 36.5 MgC ha^{-1} yr^{-1} (10 gC m^{-2} d^{-1}), which is well above woody biomass energy crops. With the harvesting penalty (80% efficiency of energy conversion) and maximum capture efficiency (90% for BIGCC with CCS) this could translate to a CDR flux $\sim$25 MgC ha^{-1} yr^{-1}. Thus with an area of 0.1 Gha (10^6 km^2) one might imagine capturing 2.5 PgC yr^{-1}, but the corresponding nutrient demand would be huge; with classic "Redfield" ratio (C : P = 106 : 1) and no recycling, it would require $\sim 2.9 \times 10^{12}$ mol P yr^{-1}, which is nearly an order of magnitude more than the global fertiliser inputs to the land of 0.39–0.45$\times 10^{12}$ mol P yr^{-1}.

To get a more reasonable upper limit on algal-BECCS CDR potential, consider the global flux of phosphorus in treated or untreated sewage of 0.048$\times 10^{12}$ mol P yr^{-1} (1.5×10^{12} gP yr^{-1}). If we imagine that this could all be linked through wastewater processing plants to fuel micro-algal productivity, with the above conversion efficiencies and no nutrient recycling, the resulting CDR flux would be 0.044 PgC yr^{-1} on a relatively modest area of 1.7 Mha. Experience with anaerobic digestion of algal biomass however suggests a nutrient recycling efficiency of 60–80% can be achieved (for nitrogen) at an energy conversion efficiency of 75%,[54] in which case the CDR flux might be increased to 0.1–0.2 PgC yr^{-1}. Still the CDR potential remains modest because of the low C : P stoichiometry of micro-algae and the assumed constraint on nutrient input. The genetic-engineering of micro-algae that excrete carbon rich (and nutrient poor) compounds such as long-chain hydrocarbon fuels may improve the forecast, but such liquid fuels are likely to be used in the transport sector where it is difficult to link them to carbon capture and storage technology.

Macro-algae cultivation has achieved productivity of 10–34 MgC ha^{-1} yr^{-1}, with much higher C : N : P stoichiometry than micro-algae.[53] Taking the "Atkinson" ratio of C : P = 550 : 1 and the global sewage flux of phosphorus,[55] with the same energy conversion and carbon capture efficiencies, a CDR flux of 0.23 PgC yr^{-1} could be generated without any nutrient recycling, although this would require a larger area than micro-algae of ~ 9–32 Mha (depending on productivity). Existing anaerobic digestion systems have shown a nutrient recycling efficiency of 60–80% can be achieved with 75% efficient energy conversion,[54,56] thus the CDR flux could conceivably be increased to 0.53–1.07 PgC yr^{-1} (on the same area). Such algaculture systems would likely be operated in estuarine (total global area ~ 100 Mha) or coastal shelf sea settings, so the required area could pose a constraint on scaling up of this technology.

Looking ahead, if the human sewage flux of phosphorus scales with total projected waste phosphorus fluxes we can expect a $\sim 50\%$ increase in 2050 and a $\sim 100\%$ increase in 2100. Thus the corresponding algal BECCS CDR flux could be ~ 0.15–0.3 PgC yr^{-1} for the micro-algal route and ~ 0.75–1.5 PgC yr^{-1} for the macro-algal route in 2050, and ~ 0.2–0.4 PgC yr^{-1} or ~ 1–2 PgC yr^{-1} in 2100. Of course if one assumed the same nutrient inputs as woody biomass energy crops grown on land then these CDR fluxes could be considerably increased.

3.3 Ocean Fertilisation

The phenomenal area of the ocean offers obvious algal-CDR potential, if carbon in biomass can be transferred to long-lived reservoirs, but once again the required nutrient inputs will be massive, and without a bioenergy gain the economics are more prohibitive. Sequestration of carbon can occur in two main reservoirs: marine sediments or the deep ocean itself. Ocean fertilisation proposals focus on adding new nutrient to the surface ocean (or increasing nutrient supply from depth), as this ultimately controls the sequestration flux of carbon. Candidate fertilisers are the macro-nutrients nitrogen (N) and phosphorus (P), and the micro-nutrient iron (Fe), noting that relieving a deficit of one nutrient is likely to lead to limitation by another. In addition to N, P and Fe limitation, silicate (Si) limitation is a possibility, especially for diatoms.

In the surface ocean, carbon fixed by photosynthesising organisms comes from dissolved inorganic carbon (DIC) in the water, creating a deficit of DIC that in turn drives an air–sea flux of CO_2. Much of the resulting organic carbon is recycled back to DIC by heterotrophic organisms in the upper ocean, nullifying the effect on atmospheric CO_2. However, a modest fraction of the carbon fixed by marine phytoplankton escapes respiration and sinks to greater depth. In coastal and shelf seas it soon hits the bottom sediments, whereas in the open ocean it can keep sinking for kilometres. However, as it does so, the sinking flux of carbon decays away due to remineralisation, following a power law function with depth,[57] with the power known to vary significantly with surface community structure.[58] The "export flux" refers to the flux of carbon sinking out of the sunlit photic zone. This has been estimated at 17 PgC yr^{-1} globally at 75 m,[59] but the bottom of the photic zone is more typically taken to be 100 m where 11 PgC yr^{-1} has been estimated.[60,61] What is critical for long-term carbon storage in the deep ocean is the "sequestration flux" below the depth of winter wind-driven mixing, which ranges over 200–1000 m, depending on location.[62] Any organic carbon remineralised to DIC above the depth of winter mixing will not create a DIC deficit on the annual or longer timescale and hence will not drive a CO_2 sink. Taking 500 m as a reference depth, there the global sequestration flux is estimated to be in the range 2.3–5.5 PgC yr^{-1} (*i.e.* 20–50% of the export flux at 100 m),[58] with a long-favoured formula giving 2.8 PgC yr^{-1} (*i.e.* 25% of the export flux).[57]

Some algal-CDR is already (inadvertently) occurring in the ocean, because humans mine and add phosphorus to the land surface 0.39–0.45×10^{12} mol P yr^{-1} and much of this leaks to the ocean. The riverine flux of biologically-available (dissolved and particulate) phosphorus to coastal seas (including sewage and detergent) has increased by 0.31×10^{12} mol P yr^{-1}, suggesting a 70–80% transfer efficiency.[63] Some of this P input is converted to organic carbon in coastal and shelf seas and buried in their sediments with a typical burial ratio under oxic (where oxygen is present) bottom waters of C : P ~ 250.[64] In the North Sea this sediment burial flux of P is around 20% of the estimated anthropogenic P input.[65] Extrapolating these figures to global coastal regions gives an estimated 0.18 PgC yr^{-1} being sequestered in their sediments at present. If the remaining 80% of anthropogenic P inputs are transferred to the open ocean and go to fuel export production there, assuming (as an upper limit) half of this flux contributes to long-term sequestration below 500 m, then a further sink of 0.16 PgC yr^{-1} is being generated at present, giving a total of 0.34 PgC yr^{-1}. Alternatively, if all anthropogenic P loading ends up in the open ocean (*i.e.* none is buried in coastal and shelf sea sediments) then the estimated sequestration flux is 0.20 PgC yr^{-1} at present.

It has been suggested that mined phosphate could be directly added to the surface ocean,[62] despite this finite resource being essential for food production. More likely we can expect inadvertent phosphorus addition to the ocean to increase in future. One projection gives a linear increase to 0.42×10^{12} mol P yr^{-1} in 2035,[63] which could drive a sink of 0.27–0.47 PgC yr^{-1}. Extrapolating the linear trend forwards gives 0.47×10^{12} mol P yr^{-1} added to the ocean in 2050 driving 0.30–0.53 PgC yr^{-1}, and 0.64×10^{12} mol P yr^{-1} added to the ocean in 2100 driving 0.41–0.71 PgC yr^{-1}. This assumes that nitrogen fixation will cause nitrogen availability in the ocean to track increased phosphorus availability,[64,66] and that micro-nutrients (*e.g.* iron) do not limit new production in the (currently coastal) regions to which phosphate is added. On the millennial timescale, the total reservoir of mineable phosphate of $323–645 \times 10^{12}$ mol P could readily be drained, which of course would precipitate a food crisis. Instead societies are going to be compelled to develop efficient recycling systems for phosphorus on land, which in turn will reduce algal-CDR potential in the ocean.

Nitrogen fertilisation of the ocean is a slightly more sustainable proposition in that nitrogen can be fixed from the atmosphere, but it comes at great energy expense. Potentially, the deficit of available nitrogen relative to phosphorus in the world ocean could be alleviated. This deficit averages 2.7 µmol kg^{-1},[67] compared to an average deep ocean nitrate concentration of 30.9 µmol kg^{-1}. Thus, removing the nitrogen deficit would result in a $\sim$9% (2.7/30.9) increase in the export flux, corresponding to about 1 PgC yr^{-1} at 100 m depth. However, it is the sequestration flux below $\sim$500 m depth which is critical and an upper limit for the increase in that is 0.5 PgC yr^{-1}. Assuming this could be achieved in 2100 on a linear growth trajectory, the 2050 CDR flux might be $\sim$0.2 PgC yr^{-1}.

Iron is the favoured nutrient for ocean fertilisation because often the C : Fe > 10000 in algal biomass, so the required nutrient inputs are much less massive than for N or P. However, iron is only limiting in remote high-nutrient low-chlorophyll (HNLC) areas of the world ocean,[68–70] notably the Southern Ocean. A number of model studies have assessed the potential carbon sink that could be generated by iron fertilisation. The maximum potential is indicated by simulations that remove iron limitation globally for 100 years.[68] Global export production across 100 m is increased initially by 3.5 PgC yr^{-1}, decaying after 100 years to 1.8 PgC yr^{-1}, and totalling 226 PgC. Diatoms are predicted to make a greater contribution to export production, creating fast sinking particles that should maximise the sequestration flux. Taking the predicted time mean increase in export flux across 100 m of 2.26 PgC yr^{-1}, and likening the remineralisation with depth to that at a station where diatoms dominate (K2 in the Northwest Pacific),[58] then the estimated increase in sequestration flux across 500 m is 1 PgC yr^{-1}. Assuming a linear growth trajectory, ~ 0.4 PgC yr^{-1} might be achieved in 2050 and 1 PgC yr^{-1} in 2100.

3.4 Combined CDR Potential

The combined potential of algal-based CDR (see Table 2) is clearly set by nutrient supply fluxes, and in the case of phosphorus and nitrogen these

Table 2 Estimates of global algal-CDR flux potential by pathway for present, 2050 and 2100.

Carbon source	Carbon store	CDR flux (PgC yr^{-1})	Key conditions / assumptions	Reference
Present				
Micro-algae	BECCS	0.044	0.048×10^{12} mol P yr^{-1} sewage	This study
	BECCS	0.1–0.2	60–80% efficient P recycling	This study
Macro-algae	BECCS	0.23	0.048×10^{12} mol P yr^{-1} sewage	This study
	BECCS	0.53–1.07	60–80% efficient P recycling	This study
River excess P flux	Sediment	0.18	C/P = 250, 0.31×10^{12} mol P yr^{-1}	32
	DIC	0.16	80% of excess P to open ocean	32
All	Various	**0.9–1.4**	Total potential	
2050				
Micro-algae	BECCS	0.15–0.3	50% increase in sewage P, recycling	This study
Macro-algae	BECCS	0.75–1.5	50% increase in sewage P, recycling	This study
River excess P flux	Sed.+DIC	0.30–0.53	0.47×10^{12} mol P yr^{-1}	32
N fertilisation	DIC	0.2	Remove 40% of global N deficit	32
Fe fertilisation	DIC	0.4	Remove 40% of global Fe deficit	32,68
All	Various	**1.6–2.6**	Total potential	
2100				
Micro-algae	BECCS	0.2–0.4	100% increase sewage P, recycling	This study
Macro-algae	BECCS	1.0–2.0	100% increase sewage P, recycling	This study
River excess P flux	Sed.+DIC	0.41–0.71	0.64×10^{12} mol P yr^{-1}	32
N fertilisation	DIC	0.5	Remove 100% global N deficit	32
Fe fertilisation	DIC	1	Remove 100% of global Fe deficit	32,68
All	Various	**2.9–4.2**	Total potential	

are the same nutrients that we need to support intensive agriculture – a function that is obviously of greater direct value to us. Thus, the potential for N and P based algal-CDR is likely to be built on the waste products of intensive agricultural systems – and will be determined by future agricultural trends. The combined, current potential of wastewater P based algal BECCS and inadvertent ocean P fertilisation appears to be significant at 0.9–1.4 PgC yr^{-1}. Conceivably this could rise to 1–2 PgC yr^{-1} in 2050, and 1.4–2.7 PgC yr^{-1} in 2100. To this might be added ocean nitrogen and iron fertilisation, to give a total algal CDR flux potential of 1.6–2.6 PgC yr^{-1} in 2050, and 2.9–4.2 PgC yr^{-1} in 2100. The present total potential is broadly comparable with plant-based CDR, and although the future potential appears smaller, this is because the woody biomass energy crop estimates implicitly include much larger nutrient inputs than allowed here for macroalgal production.

4   Alkalinity-based CDR

"Alkalinity-based CDR" refers here to all pathways in which CO_2 is removed from the atmosphere using a source of alkalinity (see Figure 1) and stored either as charge-balanced solution in seawater, or as liquid CO_2 in geologic formations (with regeneration of the alkali). It thus groups together methods of engineered direct air capture (DAC), enhanced weathering on land, and addition of alkalinity to the oceans. For methods of enhancing weathering and direct addition of alkalinity to the ocean, the key limiting resource is the supply flux of alkaline material, *i.e.* basic minerals, and the associated energy required to mine, crush and transport it. The CDR flux in turn may be limited by the kinetics of reaction of the alkali source with carbonic acid. For direct air capture (DAC) the key limiting resources are the energy required to capture CO_2 and the supply of engineered DAC devices, because the chemical sorbent system itself can be regenerated.

4.1   *Enhanced Weathering – Land*

Enhanced weathering refers to a suite of CDR options whereby a new source of alkalinity, from the dissolution of either carbonate or silicate minerals, is added to the land surface or to the oceans. Both silicate and carbonate weathering act to transfer excess CO_2 from the atmosphere to the ocean, and silicate weathering (followed by carbonate deposition) also ultimately acts to remove CO_2 to the Earth's crust. The central issue with all of these methods is that a greater mass of rock is needed than the mass of carbon removed from the atmosphere, thus the scale of the extractive industry needed to produce a useful CDR flux has to be comparable to the fossil fuel industry causing the problem in the first place.

Original suggestions for enhancing silicate weathering on land involve adding crushed olivine to soils,[71,72] particularly in the moist tropics where there are olivine sources and the reaction kinetics will be most favourable.[72,73] However, even saturating the Amazon and Congo basins with 2.2 Pg olivine yr^{-1} would only produce a CDR flux of 0.6 PgC yr^{-1}, and the efficiency penalty of crushing and milling rock is estimated at 17% making the net flux ~ 0.5 PgC yr^{-1}.[74] This requires loadings everywhere of ~ 3 kg m^{-2} yr^{-1} which may be unachievable in remote tropical locations, thus the stated upper limit potential of ~ 1 PgC yr^{-1} should be treated with caution. Also, some studies are internally inconsistent in that they show the impact on atmospheric CO_2 of larger olivine weathering CDR fluxes than they calculate to be feasible.[73,75]

4.2 Enhanced Weathering – Ocean

The weathering process can be short-circuited by adding alkalinity directly to the ocean. Proposals include adding crushed olivine directly to coastal regions[74] or to the open ocean.[76] The problem in both cases is that olivine dissolution proceeds slower under the relatively high pH of seawater and therefore smaller particle sizes are required to achieve a given kinetic dissolution rate. The idea behind coastal addition of olivine is to use the energy of waves to break down particles to smaller sizes, thus increasing the surface area for dissolution. However, the available barge fleet probably limits the CDR flux potential to an estimated 0.09–0.16 PgC yr^{-1}, which with an energy penalty of crushing and milling of $\sim 17\%$ would be reduced to 0.08–0.14 PgC yr^{-1}.[74] Open ocean olivine addition requires even smaller particles of ~ 1 μm or they will sink before dissolution is complete, increasing the energy penalty to $\sim 30\%$.[76] With a dedicated fleet of ships or addition to the ballast water of commercial ships, the resulting CDR potential[76] is 0.18–0.20 PgC yr^{-1}.

An alternative proposal – sometimes termed "ocean liming" – is to increase the alkalinity of the ocean by adding either calcium bicarbonate[77] or calcium hydroxide,[78] thus increasing CO_2 uptake. A detailed account has been presented based on using a flotilla of ships to sprinkle finely ground limestone ($CaCO_3$) on areas of the surface ocean where the depth of the saturation horizon is shallow (250–500 m) and the upwelling velocity is large (30–300 m yr^{-1}).[79] A CDR flux of 0.27 PgC yr^{-1} has been calculated after a century of linearly ramping up activity.

4.3 Direct Air Capture (DAC)

Direct air capture (DAC) refers to chemical (and physical) methods of removing CO_2 directly from the atmosphere, followed by regeneration and CO_2 storage. There are two main methods of direct air capture being developed – using a solid sorbent system (*e.g.* solid amines) or an alkaline aqueous solution (*e.g.* sodium hydroxide). The capture step is generally more expensive with solid sorbents, whereas the regeneration step is more expensive with alkaline aqueous solutions. Storage of CO_2 could be in liquid form in geological reservoirs, where ultimately the amount of CO_2 sequestered may be limited by the size of these reservoirs.[80,81] Alternatively, DAC could be combined with enhanced weathering to produce a charge neutral solution to add to the ocean, but with the additional energy cost of mining, crushing and milling the required alkaline rock material.

In principle, DAC could generate whatever size of carbon sink societies were willing to pay for, as it is unlikely to be limited by available substrates or land surface area.[13] However, the rate of development and production of DAC devices could place a serious constraint on the CDR potential in the short-medium term. One study suggests ~ 3 PgC yr^{-1} might be achievable on a 2030–2050 timeframe.[82] However, the crucial constraint on DAC is probably its high cost, which will mean other cheaper methods of CDR will be deployed before it. When considered in integrated assessment models, DAC is generally not be deployed until later this century, with one model predicting a rise from 0 PgC yr^{-1} DAC in 2065 to ~ 10 PgC yr^{-1} DAC by 2100.[83]

4.4   Combined CDR Potential

There is a shortage of studies from which to attempt alkalinity-based CDR flux estimates on different time horizons, so just their ultimate potential (assumed realisable by 2100) is considered here (see Table 3). A combination of land and ocean enhanced weathering methods might achieve 1–1.6 PgC yr^{-1}, which is comparable to the current potential of either plant-based or algal-based CDR, but much less than their ultimate potential. Direct air capture, in contrast, is the wildcard CDR technology in that its physical potential could be > 10 PgC yr^{-1} and the real constraints on its use will be economic, social and technical.

Table 3   Estimates of global alkalinity-CDR flux potential by pathway.

Process	Carbon store	CDR flux (PgC yr^{-1})	Key conditions / assumptions	Reference
Enhanced weathering				
Terrestrial olivine	DIC	0.5–1	Amazon+Congo, global	73
Coastal olivine	DIC	0.08–0.14	Constrained by barge fleet	74
Open ocean olivine	DIC	0.18–0.20	Dedicated fleet or commercial ships	76
Ocean liming	DIC	0.27	Dedicated ship fleet	79
All	DIC	**1–1.6**	Total potential	

5 Overall CDR Flux Potential

In estimating the overall physical flux potential for CDR, care must be taken to avoid double counting, when the same resource (*e.g.* land or nutrient) is required by different technologies. Competition for land (or biomass) among plant-based CDR methods has been considered above. I have also tried to avoid double-counting phosphorus supplies by only allowing algal-based CDR to access waste flows of phosphorus, either in the form of sewage through wastewater treatment, or washing off the land directly into the ocean. However, a broader competition for phosphorus between food production and plant- or algal-based CDR also ought to be considered, especially in a world where the price of rock phosphate has increased roughly 4-fold over 2006–2013 (with a peak increase of more than 8-fold in 2008). Nitrogen supplies are usually assumed to be unconstrained, because nitrogen can be fixed from the atmosphere, but there is a high energetic cost to doing that. Competition for freshwater supplies between food production and plant-based CDR also deserve more consideration.

Overall achievable CDR fluxes might also conceivably be constrained, on the century timescale, by the storage capacity for carbon in different forms. Various methods suggest that permanent forest plantations could store up to 150 PgC by 2100,[84,85] and up to $\sim$ 300 PgC in the longer term,[32] thus more than reversing historical cumulative carbon emissions from deforestation of $\sim$ 150 PgC.[86] Global storage capacity for biochar in cropland, grassland and abandoned land soils is estimated at $\sim$ 500 PgC (representing a $\sim$ 25% increase in the carbon content of the world's soils),[32] which would be difficult to saturate this century. Estimates of geologic storage capacity for liquid CO_2 range upwards from $\sim$ 500 PgC to $\sim$ 3000 PgC.[87] However, if global geologic storage is at the low end of this range and both CDR methods and conventional CCS on fossil fuel combustion are competing for this capacity, it could be filled up within this century. This would then limit BECCS and DAC fluxes ending in liquid CO_2 geological storage (see Figure 1). However, there is the alternative option to marry BECCS or DAC with enhanced weathering (at high CO_2 concentration of flue gases) and thus switch the form of carbon storage to charge neutral solution in seawater. In that case, the carbon storage capacity depends on the total mineable reserves of alkaline minerals such as olivine, limestone, and chalk, which are vast. The ocean is already the largest reservoir of carbon in the surface Earth system at $\sim$ 38000 PgC and as long as added DIC is balanced by added alkalinity, the ocean carbon store can continue to grow.

The situation is more subtle when nutrient fertilising the ocean and increasing the efficiency of the "biological carbon pump" – which transfers carbon from the equilibrated surface ocean and atmosphere to the deep ocean or the sediments as a mixture of particulate organic carbon, dissolved organic carbon, and DIC – but essentially this can increase the ocean and sediment carbon store by an amount that will be determined by the total

Table 4 Total physical CDR flux potential (PgC yr^{-1}) on different time horizons.

	'Present'	2050	2100
Plant-based CDR	0.75–1.5	2.3–5.75	5–20
Algal-based CDR	0.9–1.4	1.6–2.6	2.9–4.2
Alkalinity-based CDR	0	0.5–0.8 (+DAC?)	1–1.6 (+DAC)
Total CDR	1.65–2.9	4–9 (+DAC?)	9–26 (+DAC)

reservoirs of nutrient that can be added and the stoichiometry of the organic material produced. For phosphorus, the total mineable reservoir of phosphate of $323–645 \times 10^{12}$ mol P, if 70–80% ends up in the ocean, could sequester ~ 600 PgC.

To summarise, overall CDR is unlikely to be limited by storage constraints on the century timescale, but particular methods of CDR that end in geologic storage of CO_2 might be constrained.

Accepting these caveats, a summation of the estimated plant-based, algal-based and alkalinity-based CDR contributions (see Table 4) produces some indicative upper limit figures for the total physical flux potential for CDR. The total "present" potential of $\sim 1.5–3$ PgC yr^{-1} is based on known afforestation, known (mostly waste) flows of biomass and know "waste" flows of phosphorus. Part of this present CDR flux potential is already being realised; afforestation plus inadvertent phosphorus fertilisation of the ocean are generating CDR of 0.55–0.76 PgC yr^{-1}. The 2050 total CDR potential of ~ 4–9 PgC yr^{-1} is roughly three times the present potential. It includes (in the upper limit) a contribution from enhanced weathering of roughly half its ultimate potential. However, it does not include direct air capture, which is treated as a potential additional flux in 2050 and in 2100. The 2100 total CDR potential of $\sim 9–26$ PgC yr^{-1} carries the largest uncertainties, but is again roughly three times the 2050 potential.

6 Discussion

The literature review and "ball-park figure" estimates herein suggest there is considerable physical potential for carbon dioxide removal at a global scale. The already realised CDR flux (albeit somewhat inadvertent) of 0.55–0.76 PgC yr^{-1} is offsetting at least half of land-use change CO_2 emissions of 1.1 PgC yr^{-1}. The current physical CDR potential of $\sim 1.5–3$ PgC yr^{-1} is of a comparable size to the natural land or ocean carbon sinks and thus if realised could make a very valuable contribution to slowing the rate of rise of atmospheric CO_2.

The 2050 CDR potential of $\sim 4–9$ PgC yr^{-1}, if combined with substantive efforts to mitigate CO_2 emissions below present levels of ~ 10 PgC yr^{-1}, could together stabilise atmospheric CO_2 concentration. Although it is tempting to argue that stabilisation would happen even sooner because of the existence of natural CO_2 sinks, in fact just as CO_2 emission create these natural sinks, CO_2 removal will create a counterbalancing natural CO_2 source. Hence one must consider the net flux of CO_2 to or from the atmosphere, and when that net flux is zero, natural net CO_2 exchange fluxes will also tend towards zero.

By the end of the century, CDR potential of $\sim$9–26 PgC yr^{-1} (without including direct air capture) could offer substantial leverage on atmospheric CO_2 concentration. However, despite the large physical potential of CDR, it would not make economic sense to imagine that CDR could be ramped up to counterbalance growing CO_2 emissions, which following "business-as-usual" type scenarios could reach $\sim$30 PgC yr^{-1} in 2100. Instead economics would dictate that cheaper emissions reductions options would be taken before deploying generally more expensive CDR.

Indeed the actual CDR flux that can be realised for any method is always going to be less than the physical potential thanks to social, economic and engineering constraints. Integrated assessment models provide a framework for considering these constraints, especially cost. From the estimates herein, those models seem broadly right to bank on considerable future CDR potential. However, there are considerable uncertainties around the future estimates of CDR potential.

Starting with land plant-based CDR, the largest CDR potential is in BECCS pathways. However, land for woody biomass energy crops may simply not be available before mid-century. At the same time some integrated assessment models suggest that BECCS will only become affordable mid-century, producing a fortuitous correspondence in time. However, nutrient and freshwater demands for global scale biomass energy cropping are large and might end up being the true constraining factor.

Considering area, water and nutrient demands leads to the tentative conclusion that macro-algae based BECCS could be a better long-term option than woody biomass energy crops, because achievable productivity is higher, freshwater demands are eliminated, and nutrient demands may be lower thanks to more efficient recycling systems. Macro-algal-based CDR thus warrants further research.

Direct Air Capture (DAC) escapes from most resource constraints except for the need for energy. Hence it has great physical potential but is estimated to be very costly. This means it is not expected to be deployed at scale until CCS is implemented on all concentrated point sources of CO_2 and probably not until BECCS of some form has been deployed at scale for decades.

All of the CDR technologies discussed (with the exception of afforestation and inadvertent nutrient fertilisation) are to varying degrees unproven, especially in terms of the engineering and infrastructure required to scale them up. Such scaling up, if feasible, is bound to produce a social reaction that could readily prevent it happening. Thus a more comprehensive assessment should include the full range of social, economic and engineering constraints on different CDR options, alongside the scientific constraints considered here.

References

1. C. Le Quéré, M. R. Raupach, J. G. Canadell and G. Marland, Trends in the sources and sinks of carbon dioxide, *Nature Geosci.*, 2009, **2**, 831–836.
2. M. R. Allen, D. J. Frame, C. Huntingford, C. D. Jones, J. A. Lowe, M. Meinshausen and N. Meinshausen, Warming caused by cumulative carbon emissions towards the trillionth tonne, *Nature*, 2009, **458**(7242), 1163–1166.
3. N. E. Vaughan, T. M. Lenton and J. G. Shepherd, Climate change mitigation: trade-offs between delay and strength of action required, *Climatic Change*, 2009, **96**(1–2), 29–43.
4. M. Meinshausen, N. Meinshausen, W. Hare, S. C. B. Raper, K. Frieler, R. Knutti, D. J. Frame and M. R. Allen, Greenhouse-gas emission targets for limiting global warming to 2 °C, *Nature*, 2009, **458**(7242), 1158–1162.
5. A. Macintosh, Keeping warming within the 2 °C limit after Copenhagen, *Energy Policy*, 2010, **38**(6), 2964–2975.
6. N. Stern, *The Economics of Climate Change: The Stern Review,* Cambridge University Press, Cambridge, 2006.
7. C. Azar, D. J. A. Johansson and N. Mattsson, Meeting global temperature targets – the role of bioenergy with carbon capture and storage, *Environ. Res. Lett.*, 2013, **8**(3), 034004.
8. T. M. Lenton, Land and ocean carbon cycle feedback effects on global warming in a simple Earth system model, *Tellus*, 2000, **52B**(5), 1159–1188.
9. T. M. Lenton, Climate Change to the end of the Millennium, *Climatic Change*, 2006, **76**(1–2), 7–29.
10. T. M. Lenton, M. S. Williamson, N. R. Edwards, R. Marsh, A. R. Price, A. J. Ridgwell, J. G. Shepherd and S. J. Cox, Millennial timescale carbon cycle and climate change in an efficient Earth system model, *Climate Dynamics*, 2006, **26**(7–8), 687–711.
11. T. M. Lenton and N. E. Vaughan, The radiative forcing potential of different climate geoengineering options, *Atmos. Chem. Phys.*, 2009, **9**, 5539–5561.
12. Royal Society, *Geoengineering the climate: science, governance and uncertainty,* The Royal Society, London, 2009.
13. D. W. Keith, M. Ha-Doung and J. K. Stolaroff, Climate strategy with CO2 capture from the air, *Climatic Change*, 2006, **74**, 17–45.
14. K. R. Richards and C. Stokes, A review of forest carbon sequestration cost studies: a dozen years of research, *Climatic Change*, 2004, **63**, 1–48.
15. J. van Minnen, B. Strengers, B. Eickhout, R. Swart and R. Leemans, Quantifying the effectiveness of climate change mitigation through forest plantations and carbon sequestration with an integrated land-use model, *Carbon Balance Manage.*, 2008, **3**(1), 3.
16. N. Zeng, Carbon sequestration via wood burial, *Carbon Balance Manage.*, 2008, **3**(1), 1.

17. R. A. Metzger and G. Benford, Sequestering of atmospheric carbon through permanent disposal of crop residue, *Climatic Change*, 2001, **49**, 11–19.
18. S. E. Strand and G. Benford, Ocean sequestration of crop residue carbon: recycling fossil fuel carbon back to deep sediments, *Environ. Sci. Technol.*, 2009, **43**(4), 1000–1007.
19. J. Lehmann, J. Gaunt and M. Rondon, Bio-char sequestration in terrestrial ecosystems – a review, *Mitigation Adaptation Strategies Global Change*, 2006, **11**, 403–427.
20. F. Kraxner, S. Nilsson and M. Obersteiner, Negative emissions from BioEnergy use, carbon capture and sequestration (BECS) – the case of biomass production by sustainable forest management from semi-natural temperate forests, *Biomass Bioenergy*, 2003, **24**(4–5), 285–296.
21. K. Möllersten, J. Yan and R. Moreira, Potential market niches for biomass energy with CO2 capture and storage – opportunities for energy supply with negative CO2 emissions, *Biomass Bioenergy*, 2003, **25**(3), 273–285.
22. L. T. Evans, *Feeding the Ten Billion: Plants and Population Growth*, Cambridge University Press, Cambridge, 1998.
23. N. Nakicenovic and R. Swart, *IPCC Special Report on Emissions Scenarios*, Cambridge University Press, Cambridge, 2000.
24. M. Hoogwijk, A. Faaij, B. Eickhout, B. de Vries and W. Turkenburg, Potential of biomass energy out to 2100, for four IPCC SRES land-use scenarios, *Biomass Bioenergy*, 2005, **29**(4), 225–257.
25. B. J. Strengers, J. G. van Minnen and B. Eickhout, The role of carbon plantations in mitigating change: potential and costs, *Climatic Change*, 2008, **88**, 343–366.
26. T. W. R. Powell and T. M. Lenton, Future carbon dioxide removal via biomass energy constrained by agricultural efficiency and dietary trends, *Energy Environ. Sci.*, 2012, **5**, 8116–8133.
27. L. J. Smith and M. S. Torn, Ecological limits to terrestrial biological carbon dioxide removal, *Climatic Change*, 2013, **118**(1), 89–103.
28. L. Ornstein, I. Aleinov and D. Rind, Irrigated afforestation of the Sahara and Australian Outback to end global warming, *Climatic Change*, 2009, **97**(3), 409–437.
29. S. Luyssaert, E.-D. Schulze, A. Börner, A. Knohl, D. Hessenmöller, B. E. Law, P. Ciais and J. Grace, Old-growth forests as global carbon sinks, *Nature*, 2008, **455**(7210), 213–215.
30. FAO, *Global Forest Resources Assessment 2010 – Key findings*, Food and Agriculture Organization of the United Nations (FAO), 2010.
31. S. Wang, J. M. Chena, W. M. Jub, X. Fenga, M. Chenb, P. Chena and G. Yua, Carbon sinks and sources in China's forests during 1901–2001, *J. Environ. Manage.*, 2007, **85**(3), 524–537.
32. T. M. Lenton, The potential for land-based biological CO2 removal to lower future atmospheric CO2 concentration, *Carbon Manage.*, 2010, **1**(1), 145–160.

33. S. Nilsson and W. Schopfhauser, The carbon-sequestration potential of a global afforestation program, *Climatic Change*, 1995, **30**, 267–293.
34. S. Pacala and R. Socolow, Stabilization wedges: solving the climate problem for the next 50 years with current technologies, *Science*, 2004, **305**, 968–972.
35. P. Read, Biosphere carbon stock management: addressing the threat of abrupt climate change in the next few decades: an editorial essay, *Climatic Change*, 2008, **87**(3–4), 305–320.
36. P. Read and A. Parshotam, *Holistic Greenhouse Gas Management Strategy (with Reviewers' Comments and Authors' Rejoinders)*, Victoria University of Wellington, 2007.
37. G. Berndes, M. Hoogwijk and R. van den Broek, The contribution of biomass in the future global energy supply: a review of 17 studies, *Biomass Bioenergy*, 2003, **25**(1), 1–28.
38. S. H. Lamlom and R. A. Savidge, A reassessment of carbon content in wood: variation within and between 41 North American species, *Biomass Bioenergy*, 2003, **25**(4), 381–388.
39. E. M. W. Smeets, A. P. C. Faaij, I. M. Lewandowski and W. C. Turkenburg, A bottom-up assessment and review of global bioenergy potentials to 2050, 2007, *Prog. Energy Combust. Sci.*, 2007, **33**(1), 56–106.
40. V. Dornburg, D. van Vuuren, G. van de Ven, H. Langeveld, M. Meeusen, M. Banse, M. van Oorschot, J. Ros, G. J. van den Born, H. Aiking, M. London, H. Mozaffarian, P. Verweij, E. Lysen and A. Faaij, Bioenergy revisited: key factors in global potentials of bioenergy, *Energy Environ. Sci.*, 2010, **3**, 258–267.
41. F. Krausmann, K.-H. Erb, S. Gingrich, C. Lauk and H. Haberl, Global patterns of socioeconomic biomass flows in the year 2000: A comprehensive assessment of supply, consumption and constraints, *Ecol. Econom.*, 2008, **65**(3), 471–487.
42. D. Woolf, J. E. Amonette, F. A. Street-Perrott, J. Lehmann and S. Joseph, Sustainable biochar to mitigate global climate change, *Nature Commun.*, 2010, **1**, 56.
43. W. A. Kurz, G. Stinson, G. J. Rampley, C. C. Dymond and E. T. Neilson, Risk of natural disturbances makes future contribution of Canada's forests to the global carbon cycle highly uncertain, *Proc. Natl. Acad. Sci. U. S. A.*, 2008, **105**(5), 1551–1555.
44. K. E. Skog and G. A. Nicholson, Carbon sequestration in wood and paper products, in *The Impact of Climate Change on America's Forests: A Technical Document supporting The 2000 USDA Forest Service RPA Assessment*, ed. L. A. Joyce and R. Birdsey, Department of Agriculture, Forest Service, Rocky Mountain Research Station, Fort Collins, CO, USA, Vol. Gen. Tech. Rep. RMRS-GTR–59, 2000, pp. 79–88.
45. C. Azar, K. Lindgren, E. Larson and K. Möllersten, Carbon capture and storage from fossil fuels and biomass – costs and potential role in stabilizing the atmosphere, *Climatic Change*, 2006, **74**(1), 47–79.

46. D. Klein, N. Bauer, B. Bodirsky, J. P. Dietrich and A. Popp, Bio-IGCC with CCS as a long-term mitigation option in a coupled energy-system and land-use model, *Energy Proced.*, 2011, **4**(0), 2933–2940.
47. J. S. Rhodes and D. W. Keith, Engineering economic analysis of biomass IGCC with carbon capture and storage, *Biomass Bioenergy*, 2005, **29**(6), 440–450.
48. P. Luckow, M. A. Wise, J. J. Dooley and S. H. Kim, Large-scale utilization of biomass energy and carbon dioxide capture and storage in the transport and electricity sectors under stringent CO2 concentration limit scenarios, *Int. J. Greenhouse Gas Control*, 2010, **4**(5), 865–877.
49. S. Shackley, S. Sohi, R. Ibarrola, J. Hammond and O. Mašek, Biochar as a tool for climate change mitigation and soil management, in *Encyclopedia of Sustainability Science and Technology*, ed. Robert A. Meyers, Springer, 2012, 845–893.
50. K. G. Roberts, B. A. Gloy, S. Joseph, N. R. Scott and J. Lehmann, Life cycle assessment of biochar systems: estimating the energetic, economic, and climate change potential, *Environ. Sci. Technol.*, 2010, **44**(2), 827–833.
51. C. Lauk and K.-H. Erb, Biomass consumed in anthropogenic vegetation fires: global patterns and processes, *Ecol. Econom.*, 2009, **69**(2), 301–309.
52. D. P. van Vuuren, S. Deetman, J. Vliet, M. Berg, B. J. Ruijven and B. Koelbl, The role of negative CO2 emissions for reaching 2 °C – insights from integrated assessment modelling, *Clim. Change*, 2013, **118**(1), 15–27.
53. K. Gao and K. McKinley, Use of macroalgae for marine biomass production and CO2 remediation: a review, *J. Appl. Phycol.*, 1994, **6**(1), 45–60.
54. M. D. Hanisak, Recycling the residues from anaerobic digesters as a nutrient source for seaweed growth, *Bot. Mar.*, 1981, **24**(1), 57–62.
55. M. J. Atkinson and S. V. Smith, C:N:P ratios of benthic marine plants, *Limnol. Oceanogr.*, 1983, **28**(3), 568–574.
56. A. d. R. N'Yeurt, D. P. Chynoweth, M. E. Capron, J. R. Stewart and M. A. Hasan, Negative carbon via Ocean Afforestation, *Process Safety Environ. Protect.*, 2012, **90**(6), 467–474.
57. J. H. Martin, G. A. Knauer, D. M. Karl and W. W. Broenkow, VERTEX: carbon cycling in the northeast Pacific, *Deep Sea Res. A*, 1987, **34**(2), 267–285.
58. K. O. Buesseler, C. H. Lamborg, P. W. Boyd, P. J. Lam, T. W. Trull, R. R. Bidigare, J. K. B. Bishop, K. L. Casciotti, F. Dehairs, M. Elskens, M. Honda, D. M. Karl, D. A. Siegel, M. W. Silver, D. K. Steinberg, J. Valdes, B. Van Mooy and S. Wilson, Revisiting carbon flux through the ocean's twilight zone, *Science*, 2007, **316**, 567–570.
59. R. G. Najjar, X. Jin, F. Louanchi, O. Aumont, K. Caldeira, S. C. Doney, J.-C. Dutay, M. Follows, N. Gruber, F. Joos, K. Lindsay, E. Maier-Reimer, R. J. Matear, K. Matsumoto, P. Monfray, A. Mouchet, J. C. Orr, G.-K. Plattner, J. L. Sarmiento, R. Schlitzer, R. D. Slater, M.-F. Weirig, Y. Yamanaka and A. Yool, Impact of circulation on export production,

dissolved organic matter, and dissolved oxygen in the ocean: results from Phase II of the Ocean Carbon-cycle Model Intercomparison Project (OCMIP-2), *Global Biogeochem. Cycles*, 2007, **21**, GB3007.

60. E. A. Laws, P. G. Falkowski, W. O. Smith, H. Ducklow and J. J. McCarthy, Temperature effects on export production in the open ocean, *Global Biogeochem. Cycles*, 2000, **14**(4), 1231–1246.

61. M. Gehlen, L. Bopp, N. Emprin, O. Aumont, C. Heinze and O. Ragueneau, Reconciling surface ocean productivity, export fluxes and sediment composition in a global biogeochemical ocean model, *Biogeosciences*, 2006, **3**, 521–537.

62. R. S. Lampitt, E. P. Achterberg, T. R. Anderson, J. A. Hughes, M. D. Iglesias-Rodriguez, B. A. Kelly-Gerreyn, M. Lucas, E. E. Popova, R. Sanders, J. G. Shepherd, D. Smythe-Wright and A. Yool, Ocean fertilization: a potential means of geoengineering, *Philos. Trans. R. Soc., Ser. A*, 2008, **366**(1882), 3919–3945.

63. F. T. Mackenzie, L. M. Ver and A. Lerman, Century-scale nitrogen and phosphorus controls of the carbon cycle, *Chem. Geol.*, 2002, **190**, 13–32.

64. T. M. Lenton and A. J. Watson, Redfield revisited: 1. Regulation of nitrate, phosphate and oxygen in the ocean, *Global Biogeochem. Cycles*, 2000a, **14**(1), 225–248.

65. N. Brion, W. Baeyens, S. De Galan, M. Elskens and R. W. P. M. Laane, The North Sea: source or sink for nitrogen and phosphorus to the Atlantic Ocean?, *Biogeochemistry*, 2004, **68**(277–296).

66. A. C. Redfield, The biological control of chemical factors in the environment, *Am. Sci.*, 1958, **46**, 205–221.

67. L. A. Anderson and J. L. Sarmiento, Redfield ratios of remineralization determined by nutrient data analysis, *Global Biogeochem. Cycles*, 1994, **8**(1), 65–80.

68. O. Aumount and L. Bopp, Globalizing results from ocean in situ iron fertilization studies, *Global Biogeochem. Cycles*, 2006, **20**, GB2017.

69. X. Jin, N. Gruber, H. Frenzel, S. C. Dooley and J. C. McWilliams, The impact on atmospheric CO_2 of iron fertilization induced changes in the ocean's biological pump, *Biogeosciences*, 2008, **5**, 385–406.

70. R. E. Zeebe and D. Archer, Feasibility of ocean fertilization and its impact on future atmospheric CO_2 levels, *Geophys. Res. Lett.*, 2005, **32**, L09703.

71. T. Kojima, A. Nagamine, N. Ueno and S. Uemiya, Absorption and fixation of carbon dioxide by rock weathering, *Energy Conver. Manage.*, 1997, **38**(0), S461–S466.

72. R. D. Schuiling and P. Krijgsman, Enhanced weathering: an effective and cheap tool to sequester CO_2, *Clim. Change*, 2006, **74**, 349–354.

73. P. Kohler, J. Hartmann and D. A. Wolf-Gladrow, Geoengineering potential of artificially enhanced silicate weathering of olivine, *Proc. Natl. Acad. Sci.*, 2010, **107**(47), 20228–33.

74. S. J. T. Hangx and C. J. Spiers, Coastal spreading of olivine to control atmospheric CO_2 concentrations: a critical analysis of viability, *Int. J. Greenhouse Gas Control*, 2009, **3**(6), 757–767.

75. J. Hartmann, A. J. West, P. Renforth, P. Köhler, C. L. De La Rocha, D. A. Wolf-Gladrow, H. H. Dürr and J. Scheffran, Enhanced chemical weathering as a geoengineering strategy to reduce atmospheric carbon dioxide, supply nutrients, and mitigate ocean acidification, *Rev. Geophys.*, 2013, **51**(2), 113–149.

76. P. Kohler, J. F. Abrams, C. Volker, J. Hauck and D. A. Wolf-Gladrow, Geoengineering impact of open ocean dissolution of olivine on atmospheric CO2, surface ocean pH and marine biology, *Environ. Res. Lett.*, 2013, **8**(1), 014009.

77. G. H. Rau and K. Caldeira, Enhanced carbonate dissolution: a means of sequestering waste CO2 as ocean bicarbonate, *Energy Conversion Manage.*, 1999, **40**(17), 1803–1813.

78. H. S. Kheshgi, Sequestering atmospheric carbon dioxide by increasing ocean alkalinity, *Energy*, 1995, **20**(9), 915–922.

79. L. D. D. Harvey, Mitigating the atmospheric CO2 increase and ocean acidification by adding limestone powder to upwelling regions, *J. Geophys. Res. (Oceans)*, 2008, **113**, C04028.

80. IPCC, *Carbon Dioxide Capture and Storage,* Cambridge University Press, Cambridge, 2005.

81. K. Z. House, D. P. Schrag, C. F. Harvey and K. S. Lackner, Permanent carbon dioxide storage in deep-sea sediments, *Proceed. Natl. Acad. Sci. U. S. A.*, 2006, **103**(33), 12291–12295.

82. D. McLaren, A comparative global assessment of potential negative emissions technologies, *Process Safety Environ. Protect.*, 2012, **90**(6), 489–500.

83. C. Chen and M. Tavoni, Direct air capture of CO2 and climate stabilization: a model based assessment, *Clim. Change*, 2013, **118**(1), 59–72.

84. J. Sathaye, W. Makundi, L. Dale, P. Chan and K. Andrasko, GHG mitigation potential, costs and benefits in global forests: a dynamic partial equilibrium approach, *Energy J.*, 2006, Special Issue 3: Multi-Greenhouse Gas Mitigation and Climate Policy, 127.

85. B. Sohngen and R. Sedjo, Carbon sequestration in global forests under different carbon price regimes, *Energy J.*, 2006, Special Issue 3: Multi-Greenhouse Gas Mitigation and Climate Policy, 109.

86. J. G. Canadell and M. R. Raupach, Managing forests for climate change mitigation, *Science*, 2008, **320**(5882), 1456–1457.

87. B. Metz, O. Davidson, H. D. Coninck, M. Loos and L. Meyer, *IPCC Special Report on Carbon Dioxide Capture and Storage,* Cambridge University Press, Cambridge, 2005.

The Use of Artificial Trees

KLAUS S. LACKNER

ABSTRACT

Direct capture of carbon dioxide from ambient air with devices that resemble trees could contribute to a net zero carbon economy and even support a level of negative emissions sufficient to drive the concentration of carbon dioxide in the atmosphere in a matter of decades back down to acceptable levels. Direct air capture adds a new capture method to the carbon capture and storage technology suite. It can work with all storage options and cancel out emissions from any source. Point sources, in the main, would be better off capturing their own emissions instead of releasing them to the atmosphere. Capture from air would likely focus on emissions from the transportation sector. Here, air capture can also support a closed carbon cycle that starts with carbon dioxide from the air and non-fossil energy and produces liquid fuels which, after use, return their carbon back to the atmosphere. Air capture can retrieve carbon dioxide that has been released to the air in the past, and thus reverse emissions and limit their damage to the duration they were allowed to reside in the environment. The ability to reverse emissions adds a new dimension to policy options.

1 Introduction

Without ready access to energy it would be impossible to provide seven billion people with food, water and raw materials. Energy also supports manufacturing, information processing and transportation. Even the reduction of the environmental footprint of human activities requires energy. Furthermore, the best hope for a sustainable future is to achieve,

Issues in Environmental Science and Technology, 38
Geoengineering of the Climate System
Edited by R.E. Hester and R.M. Harrison

across the globe, the minimum level of prosperity at which population growth slows and eventually stops. Just reaching this level would likely drive energy consumption higher than it is today, but economic growth is unlikely to stop once this minimal requirement has been met, and energy demand can be expected to grow well beyond this minimum.

Unfortunately, current energy infrastructures pose one of the largest obstacles to sustainable development. Most energy is derived from fossil fuels by methods that ignore the environmental problems associated with the carbon dioxide released during fuel combustion. As long as this effluent is discharged into the atmosphere, the carbon dioxide concentration in the air will continue to rise and the damage will persist for a thousand years.[1] Somehow, the world will have to stabilize this concentration, possibly below today's level. To this end, it has been suggested that artificial trees could help stabilize or even reduce the carbon dioxide concentration in the air.[2–5]

In this context, the term "artificial tree" does not refer to a winter holiday decoration,[†] but to a machine that extracts carbon dioxide directly from ambient air and delivers it in a more concentrated stream. Depending on the application, this concentration could range from a few thousand parts per million added to air, all the way to pure carbon dioxide at high pressure. The analogy to a tree was motivated by the passive nature of a device with wind flowing unimpeded over surfaces that, just like the leaves of a tree, absorb carbon dioxide. The analogy breaks down at photosynthesis, because the machine is supposed to produce more concentrated carbon dioxide, but not to convert it to a more energetic compound. In short, the artificial tree is a capture device that takes carbon dioxide from ambient air. The process is sometimes referred to as direct air capture or DAC.

Air capture of carbon dioxide does not necessarily involve machinery or collectors. Other methods for removing carbon dioxide from the air have been proposed but differ from direct air capture that uses machinery. For example, accelerating the growth of biomass on land or in the ocean would also remove carbon dioxide from air,[6–8] as would adding alkalinity to the ocean.[9,10] In both cases, the removal of carbon dioxide from the air can be a spontaneous process that does not require direct air capture. Nevertheless, growing algae with carbon dioxide obtained by direct air capture,[11] or reacting alkalinity with directly captured carbon dioxide have been considered.[12]

Why would artificial trees be useful and would they be economically viable? These two questions cannot be separated. It is certainly possible to collect carbon dioxide from air. It has been done in a number of different circumstances ranging from air cleanup in submarines to the removal of carbon dioxide from air that is to be cryogenically liquefied. Instead one has to ask, whether the value of the application can support the cost of the process. Unfortunately, the future cost of a process still under development is typically unknown and will have to be discovered by trial and error.

[†]To avoid this confusion I initially tried to use the term "synthetic tree," but it does not seem to resonate.

Experience from other fields shows that there is no shortcut for learning.[13] Initial costs are likely to be high, but costs are also likely to come down over time. In many fields, costs dropped tenfold and even hundredfold over the course of decades. Success may depend on the availability of small market niches in which the perceived value is high and the technology can be tried out.

I will begin by summarizing the challenges that air capture will have to overcome and move on to sketch out a path to the solution with an unavoidable emphasis on our own approach. Then I will discuss the potential uses of air capture and its implications for policy and technology development.

2   Air Capture as an Engineering and Policy Challenge

Artificial trees and air capture seem to stir emotional responses from advocates and detractors. The concept seems to hit a nerve, most likely because it does not fit into ready-made categories. It challenges preconceived notions of what can be done and, for some people, it even raises the question what should be done. The availability of air capture technology could change the debate on climate change and possibly allow the future use of hydrocarbon fuels, which some would prefer to see outlawed. Objections range from the perception of insurmountable technical difficulties, as for example in the APS study,[14] which seems to see little room for direct air capture, to deep-rooted suspicions against any technology that aims to fix problems with previously introduced technologies.

From a policy perspective, direct air capture, if it proves affordable, offers a number of advantages. As pointed out by Sarewitz and Nelson,[15] it is a technology that addresses the problem directly. If excess carbon dioxide is the problem, then air capture removes the problem. The result is easily measurable and detectable, so there need be little debate about how much carbon dioxide has actually been removed. While Sarewitz and Nelson wondered why the technology has not been investigated much, they assumed that given sufficient resources it could be developed. They concluded that it has the right attributes for a successful technological fix, but this, of course, depends on its technical viability, which has been questioned by skeptics, most notably a study performed by the American Physical Society.[14]

The technology skeptics should not be ignored: air capture poses severe technical challenges. First, carbon dioxide in air is quite dilute. It has been argued that the cost per unit of output of different separation technologies varies broadly linearly with dilution. This is known as Sherwood's rule.[16] Applying this rule to cost extrapolations from known gas separation processes that operate at much higher concentrations than the 400 ppm of carbon dioxide in the air would lead one to believe that air capture technology will always be too expensive.[17] It is therefore important to find ways around Sherwood's rule, which is a rule of thumb rather than a law of physics.[18] As discussed below, there are exceptions to Sherwood's rule and one can make the case that a well-designed air capture technology need not abide by it.

A second problem, less often mentioned, but equally important, is the competition of carbon dioxide and water in the separation process. Even though the water content of the air varies with temperature and humidity, it is usually much higher than the air's carbon dioxide content. The concentration of water vapor in the air on a hot day can exceed 30 000 ppm, but even on a cold day it is easily 5000 ppm, which is still an order of magnitude larger than the concentration of carbon dioxide. If water were captured at the same cost as carbon dioxide, the process could easily "drown" in expensive water. However, as shown below, there are technologies that avoid this problem.

Air capture as a tool to manage climate change would need to operate at a formidable scale. To be an important player that has impact on climate change, the technology would need to collect many gigatons of carbon dioxide per year. A figure of merit for carbon dioxide capture as capture of last resort is 1 to 10 gigatons per year. At this rate, air capture would collect some but not all the carbon dioxide currently emitted. By contrast, long term reductions in atmospheric carbon dioxide should be measured in parts per million per year and thus could require annual removals of 10 to 100 gigatons. The challenge is not only how to operate at such large scales, but also how to bootstrap a new technology and infrastructure to such scales in a reasonably short time.

3 An Example of an Air Capture Technology

These challenges are best discussed in the context of a specific implementation. Due to familiarity, I focus on the implementation we have developed.[19] This is not to say that ours is the only approach, see for example Refs. 20–23, or that ultimately there may not be better solutions, but by looking at the particulars of a specific implementation, one can show how the obstacles I laid out above might be overcome.

Although our approach involves a sorbent cycle, it is not an extrapolation of conventional gas separation technologies. It represents a new concept that directly addresses the three issues of dilution, water interference, and scaling. We have developed a passive, wind-driven design that minimizes the cost of bringing the air in contact with a sorbent that is regenerated with humidity. The process consumes rather than collects water. The contacting process relies entirely on ambient energy resources and the energy obtained from water evaporation is used to raise the carbon dioxide partial pressure hundredfold.[19] The overall design we have proposed can be mass manufactured, which offers the opportunity for large cost reductions. In looking at the aggregate scale of small mass-produced machinery, one can point to many other technologies that operate at extremely large scales. For example, the power capacity of the car engines produced in the United States in a single year is comparable to the entire standing electric power plant capacity in the US.

In summary, we rely on three technological advances to assure the viability and affordability of direct air capture: passive air contacting, moisture-driven regeneration, and scale-up by mass production.

Passive Air Contacting: we minimize the cost of the sorbent-air contactor through a passive design with very low pressure drops. A passive design gives us the means of side-stepping Sherwood's rule. The first step in any separation involves handling the entire volume of the mixture. If this step dominates overall costs, then Sherwood's rule must apply. Conversely, it appears that for most processes the cost of the initial step accounts for the majority of the cost. Therefore Sherwood's rule usually applies. For example, the cost of extracting metals from ores is essentially accounted for by the cost of mining, crushing and grinding the ore plus the cost for tailings disposal.[18] On the other hand, if the first step involving the entire volume only contributes little to the overall cost, then there is no reason why the remaining cost would scale linearly in the dilution. For example, in capturing carbon dioxide from air, the size of the collector which brings the sorbent in contact with the air scales linearly with the dilution and so will the amount of energy that is spent on transporting the air volume through the device. However, the sorbent mass is proportional to the amount of carbon dioxide that is to be bound on the sorbent, and this mass bears no relation to the original dilution and could be essentially constant with increasing dilution. Costs will still rise, because the regeneration energy increases logarithmically with dilution.[24]

There has been an analogous development in another technology, the extraction of uranium from seawater, which has been studied for decades.[25,26] Seawater with 3 ppb of uranium is brought in contact with a sorbent that selectively binds uranium. In early implementations the cost of pumping water through sorbent filters resulted in unacceptable costs. In recent designs, the sorbents have been reduced to strands of buoyant plastic anchored to the ocean floor, in effect artificial kelp.[27] The contacting process is now entirely passive. The main cost is in harvesting and processing the artificial kelp. As a result costs have been reduced three or four orders of magnitude below those implied by Sherwood's rule. The basic concept is the same as for artificial trees; passive contacting eliminates nearly all those costs that are linearly related to dilution.

Moisture Driven Regeneration: the competitive sorption of water vapor from the air on the sorbent can be avoided by saturating the sorbent with water prior to contacting air.

The sorbent we use is an anionic exchange resin that binds carbon dioxide tightly when dry, and releases it again when moist.[28,29] The equilibrium partial pressure of carbon dioxide over resin with a given carbon dioxide loading increases several hundredfold as the relative humidity moves from a few percent to nearly 100 percent. Resin equilibrated in ambient air, can release the carbon dioxide at a partial pressure of 5 to 10 kPa, simply by exposing the resin to 100 percent humidity or to liquid water. This moisture swing provides an exceedingly efficient way of raising the CO_2 partial pressure hundredfold.[‡] Up to this stage, very little energy has been consumed directly. Wind and water evaporation drive the process, with additional energy demand limited to mechanical motion of system components. Producing a concentrated and pure stream of carbon dioxide from the output of this first stage at one atmosphere or higher pressure requires significant additional energy. After this initial step, the problem is more akin to flue gas scrubbing than direct air capture. However, it is this second stage of the process that dominates cost estimates and energy demand.

Thus, in our approach the capture is staged, and the output of the first stage is a stream with a low partial pressure of carbon dioxide that can take on several forms:

(i) A stream of carbon dioxide with some water vapor and little other admixtures at a total pressure between 5 and 10 kPa (raising the temperature to about 45 °C increases pressure);

(ii) A stream of carbon dioxide enriched air (or nitrogen, or oxygen) at ambient or sub-ambient pressure with a carbon dioxide partial pressure around 5 kPa;[§] and

(iii) A clean bicarbonate solution produced by washing the loaded resin in a clean carbonate solution (the solution must be substantially free of other anions like sulfates or chlorides that otherwise it would deactivate the sorbent).[¶]

Moisture driven regeneration is energetically efficient, and in addition offers a solution to water interference. The wet stage of the system is in the regeneration not in the collection mode. As a result, the sorbent loses water when exposed to air, and one is not in danger of collecting water at high cost. Unlike other systems that also can keep the sorbent wet during regeneration, for example by using steam for a thermal swing,[30] the humidity swing actually benefits energetically from drying in the air. As the resin dries, its affinity to carbon dioxide increases rapidly. The binding of carbon dioxide is made possible by the heat of evaporation of water.

While it has been argued by comparison with other technologies that the energy demand of air capture is bound to be too high,[17] I maintain

---

[‡]The pressure boost is less than the pressure amplification at constant loading, as there needs to be room for the unloading of the resin. This loss can be reduced by counterstreaming sweep gas over a series of partially unloaded sorbent filters.

[§]Other sweep gases are possible. However, the sorbent is exposed to open air. This raises concerns over losses and air pollution from sweep gas escaping.

[¶]It is not necessary that the solution is produced from deionized water. In order to avoid exchanging carbonate and bicarbonate ions on the resin with impurity anions, the carbonate or bicarbonate concentration in the washing solution has to be much larger than that of the impurity ions. Freshwater sources are clean enough, but seawater is too rich in chloride to mask it with additional carbonate.

that these high energy demands do not follow from fundamental laws of physics, but are based on Sherwood's empirical rule. If the dominant energy cost were in the first step, then energy demands would indeed scale linearly in the dilution. Fortunately, unlike in metal extraction, there is no need for crushing and grinding. By avoiding fans and blowers, air capture arrives at an energy cost that is mainly in the regeneration of the sorbent and only scales logarithmically with dilution. I and others have shown that the total energy consumption for sorbent regeneration need not be much larger than it is for flue gas scrubbing.[24,14] Furthermore, we take advantage of the evaporation of water in air, which proves to be a very cheap way of applying energy, which conventionally is not counted in the energy balance.[‖] The increase in partial pressure achieved in the humidity swing amounts to about half of the energy required in the compression process.

Mass Production Based Scaling: we observe that large cost reductions in the commercial sector have been introduced wherever mass production methods were successful.[31] Mass production allows for an iterative process of manufacturing which learns over time how to reduce costs. Our approach to air capture aims to take advantage of the power of the learning curve, which states that costs in mass manufacturing drop with cumulative output. Every doubling in cumulative output reduces unit cost by 15 to 20 percent. To take advantage of mass production, one must deploy many small units that operate in parallel. Mass manufactured car engines represent a special implementation of a small power plant that is about one hundred times cheaper per kW of capacity than a large coal fired power plant.[31] A natural scale for air capture is set by the size of a standard shipping container. An air capture device that can fold up into a container can easily be shipped to its point of use. This makes it easy to use factory built devices while keeping the collection system at the point of carbon dioxide consumption. If carbon dioxide demand shifts to a different location, the collectors can easily follow. The container size, which in terms of mass and complexity is comparable to that of a car or truck, lends itself to mass production. The number of units required to provide the carbon dioxide for a single storage site or for a refinery producing fuel would indeed be very large.

How big a collector can be put into a standard shipping container will depend on the state of the technology. However, the size of the collector standing in the wind is largely determined by the concentration of carbon dioxide in the air. A capture of 100 ppm from air and an air flow speed through the device of 1 m s^{-1} would result in a collection of about 15 kg m^{-2} per day. Therefore it takes 65 m^2 of wind-facing surface to collect 1000 kg day^{-1}. Based on the current state of the technology, such a device

---

[‖] Drying a towel on a clothesline in the wind consumes a large amount of heat of evaporation but this energy is delivered for free by the air passing over it. The same happens to the sorbent material that in drying slightly cools the air passing over it.

could readily fit into a shipping container. At 30 cm depth for the filter structure, which appears reasonable,[32] the volume of the filters standing in the wind would be about 25 m^3, a typical container is 68 m^3. We therefore assume that a unit size would be on the order of one-ton-per-day, would fit into a standard shipping container, and would have a complexity level comparable to that of a car or a light truck. The only way to verify this assumption is to actually demonstrate it by designing and constructing a number of such devices.

With such basic units one can start the process of scaling by deploying many units in parallel. Three thousand such one-ton-per day units would collect a million tons of carbon dioxide per year, and could easily be put onto a square kilometer of land. More units on the same area would experience significant interference as downwind units would see less carbon dioxide in the air.[33,14] Collecting this much carbon dioxide, not necessarily all in one place, could deliver one eighth of the merchant** carbon dioxide produced in the United States. One hundred thousand units could roughly satisfy the current need for carbon dioxide in enhanced oil recovery operations.

10 million such units would collect 3.6 gigatons per year or about 12 percent of the current world emissions. Assuming a life time of 10 years, the required manufacturing rate would be about one million units per year. This is small compared to the current world capacity to build cars or light trucks, which is about 85 million units per year.[34]

Lastly, matching the scale of world emissions would require another order of magnitude in the number of units. At a production rate of 10 million units per year or a standing fleet of 100 million units, direct air capture would more than cancel out current carbon dioxide emissions. Even this scale is dwarfed by current manufacturing capacities. Shanghai harbor ships 30 million containers per year, servicing a manufacturing capacity much larger than necessary for air capture to become an important contributor to carbon management.

Mass production is an extremely powerful way of reducing costs. The cost of lighting dropped seven thousandfold during the 20th century.[35] Mass production not only reduces cost of individual units, but aggregate outputs can also be very large. As mentioned above, the annual production of car engines in the United States adds a power output capacity that exceeds the entire power plant fleet. This not only illustrates the size of aggregate output, but the mismatch in cost between car engines and power plants also shows the difficulty in predicting cost and future designs of any infrastructure. Appealing to the car engine analogy, one would have misjudged power plant costs by two orders of magnitude.

**Industrial carbon dioxide typically shipped by refrigerated tanker truck and sold for a wide range of applications from filling fire extinguishers to producing dry ice.

4   Cost Issues

There is no question that capture of carbon dioxide from air is possible. It is being done. The question is not whether carbon dioxide capture can be accomplished but whether it can be accomplished at a cost that makes it useful for more than small, high-value applications. We argue that this question can only be answered by developing the technology to the point that it can be tested in the real world, and pushed to the point where learning will begin to drive the costs down.

We have argued elsewhere that if the long-term cost of capture and disposal of carbon dioxide would prove to be much larger than $100 per ton of carbon dioxide, these technologies would have little practical interest in the climate change debate.[13] They would raise the effective cost of coal by about $350 per ton, the cost of natural gas by $5 per gigajoule, and the cost of oil by about $40 per barrel. If carbon regulations where to result in carbon dioxide prices in excess of $100 ton^{-1}, they would effectively destroy the competitive position of fossil fuels relative to renewable and nuclear energy sources. As the price of carbon increases, coal, gas and oil will be affected in this order. Oil may be able to hold on longer, because much of the oil on the market is lifted from the ground at much lower costs than it is sold, and thus the oil price could absorb a significant "carbon surcharge." Its unique role in the transportation sector also makes substitution harder resulting in low demand elasticity. As a consequence, oil may be an interesting target for introducing carbon constraints. European gasoline prices have enough tax built into them that the cost of carbon dioxide remediation would not look all that daunting. $100 ton^{-1} of CO_2 translates into about 25¢ liter^{-1} of gasoline.

Immediate costs of air capture are likely to be high. We will argue below that this starting price could be around $200 ton^{-1} of CO_2. If indeed true, this suggests the following questions: at $200 per ton can one find small niche markets? Can air capture drop below $100 per ton and establish a large foothold in the market? Can air capture ultimately be cost competitive with other methods of reducing carbon dioxide accumulation in the air? Companies who work on air capture technologies have all argued that they can deliver carbon dioxide below this starting price, but until they have actually done so, this claim remains untested.

Direct air capture has the ability to start small in niche markets, because unlike coal plant scrubbing it can start small. The daily demand for merchant carbon dioxide in midsized towns is measured in tons. The per capita consumption in the United States is about 25 kg per year. Prices can go as high as $300 ton^{-1} in places far from sources, as for example in Phoenix, Arizona, which receives carbon dioxide from Los Angeles. If air capture costs are a few hundred dollars per ton, the demand would be very small and limited to very pricey niche applications. Such applications include uses at remote sites that require carbon dioxide that otherwise would have to be shipped by truck. Delivering to such small markets would introduce learning, which would likely result in cost reductions. As the cost of carbon dioxide generation drops, the potential market will grow and with it the opportunity to learn more. At $100 per ton, the market is likely to be measured in tens of millions of tons, as the world market for merchant carbon dioxide is already on this scale and the typical price of industrial carbon dioxide tends to be around this level in locations that have ready access to carbon dioxide.

For carbon dioxide priced between $50 and $100 per ton, the market for enhanced oil recovery, which is potentially very large, will start to provide a strong driver for the introduction of such a new technology. This market should not be viewed as a way of reducing carbon dioxide emissions, but as a way of introducing a technology that has certain advantages over laying pipelines. First and foremost, air capture offers flexibility. If units are small and delivered by truck, then the number of units operating in a field can be matched to fluctuating demand and units can move from one site to another as demand patterns change over months or years. By choosing this technology over pipelines one does not lock in a steady supply for thirty years, but rather acknowledges the uncertainties in a rapidly changing market. Air captured carbon dioxide could sell at a risk premium making enhanced oil recovery an interesting starting point for the technology.

The price of carbon dioxide reductions to avoid climate change has not been discovered yet. The reason is simple; current schemes for carbon dioxide pricing see no demand pressure and thus their cost vary widely. If policies change and emphasis moves from environmental concerns to economic competitiveness, the carbon price can drop precipitously as has become very evident in the European Carbon Exchange. Nevertheless, as the reality of climate change sets in, the price of carbon dioxide reductions will continue to rise until a net zero carbon economy is achieved.

On the other hand, some countries, like Sweden, have implemented rather stringent carbon controls and the carbon price manifested in a tax can be as high as $200 per ton. This exceeds our long term upper limit and thus should lead to a gradual phasing out of fossil carbon, which indeed is compatible with a strong downward trend seen in Sweden.[36]

5   What Price can Air Capture Technology Deliver?

There have been a number of attempts to estimate the cost of air capture technologies. Cost estimation of new technologies is nearly impossible. At the very least, one has to consider carefully the assumptions that go into such estimates and their inherent uncertainties. From a long-term policy perspective it is not very interesting to know what a first-of-a-kind plant will cost. This becomes even less important if the cost of a first prototype is affordable, because individual prototype units are small. In a mass production approach, overall cost of a project resulting in few prototype units would likely be measured in millions rather than the billions that would be needed to demonstrate scaling up to a size that matches the output of a power plant. While not very interesting from a policy perspective, the initial cost of the technology is the only cost one might be able to estimate in a reliable fashion.

Hindsight has shown many examples of relevant technologies dramatically dropping in price after they have been introduced.[13] This occasionally applies to large scale, environmental technologies like flue gas scrubbing for sulfur compounds, which became much cheaper once it was mandated and tied to sulfur emissions trading. However, often it applies to technologies firmly rooted in mass manufacturing. Solar panels today are more than 100 times cheaper than they were in the 1960s. Televisions and cars have added performance while staying roughly constant in cost.

Future cost of technologies is not predictable, since the advances required to achieve cost reduction are not yet realized. Assuming that such advances will always materialize would also be a bad assumption; there could be an enormous selection bias in favor of technologies that are able to drive costs down. Technologies that failed may have in part failed because developers were unable to maintain the necessary rate of cost reductions. Forecasting with uncertainties that span orders of magnitude is not very useful.

On the other hand, if the APS study which estimates the price of air capture at \$600 ton^{-1} for avoiding a ton of CO_2, sits at the beginning of a long learning curve,[14] then a reduction to below \$100 per ton does not look very daunting compared to known cost reductions in other technologies. This argument is further strengthened by the observation that the APS study assumed that there is a large cost in blowing the air through packed bed collectors, that the sorbent would transfer its carbon dioxide to calcium carbonate which has to be calcined at temperatures above 700 °C, and that the calcium carbonate enters this cycle wet and thus a large amount of energy is used to simply to evaporate water. Removing these large cost items that are not present in our current design should result in large savings. In effect, compared to the process studied for the APS, which by design aimed to use off-the-shelf technologies,[37] our process has undergone a significant amount of learning and it is not at all surprising that it seems to start from a lower cost. In 2008 it has been suggested by the startup company, I have been involved with, that a first of a kind cost could be on the order of \$250 ton^{-1} of CO_2. This number, or today's lower estimates, cannot be validated except through a technology demonstration. However, in spite of the fact that the process is far simpler than the APS process, it is still nearly half of the capture cost given by the APS report.[††] Learning will start from here and past experience suggests that the ultimate lower bound on the cost of such a process could be very much lower.

For purposes of this discussion, it might be useful to introduce the concept of frictionless cost. The simple fact that costs can come down by orders

---

[††]The capture cost as opposed to the cost of avoided emission has been estimated at \$550 ton^{-1} of CO_2 by the APS.

of magnitude over time makes it clear that the initial cost of new machinery is not set by any physical limits but by the friction resulting from myriad design compromises which choose known and validated approaches over sometimes yet undiscovered approaches better suited to the new system. This frictional cost is reduced over time as the system becomes more and more streamlined and efficient. Sometimes it takes many decades before the cost of operation approaches physical limits and thus the ultimate frictionless cost. Frictionless cost accounts for energy demands that are rooted in thermodynamics and thus cannot be avoided and it includes the cost of minimum material inputs that are needed as raw materials for the process or as structural materials to hold the machinery together. We have argued in the past that this frictionless cost for air capture is below $30 per ton of carbon dioxide.[19] Whether this frictionless limit can be reached, or how long it will take to reach it, remains to be seen. In the meantime, a cost below $100 per ton would have a profound impact on the climate change debate.

From a policy perspective the concept of a frictionless cost should be of interest. If process costs trend in the long run toward their frictionless limits, then it is useful to compare different processes from this perspective. Such a comparison puts advanced and new processes on a similar footing as it compares the endpoint of both developments. However, it does highlight a large uncertainty in the cost estimate. The less developed a process is, the larger this uncertainty becomes. Applying the concept of frictionless cost to air capture, I argue that in the frictionless limit the cost of contacting the air is small compared to the cost incurred in sorbent regeneration. This makes it possible to focus on the latter. In comparison to flue gas scrubbing, one can see from energy requirements that sorbent regeneration costs for flue gas scrubbing and air capture are similar, but that air capture requires more energy. This increased energy demand suggests costs 1.5 to 2 times higher, depending on some of the underlying assumptions.[24] Noting that costs for flue gas scrubbing as low as $15 per ton have been estimated,[38] it stands to reason that similar costs in air capture could reach as low as $30 ton^{-1}.[‡‡]

The discussions of frictionless costs should not hide the fact that the immediate cost for the first practical device will be much higher. Once such a system has been designed in detail, it is perfectly possible to estimate the cost of this device or – even better – determine the actual costs that are paid to build this device. While it is true that one might hope for hard to quantify cost reductions that come with more experience in building such a device, the initial investor or government support will have to bear the cost of this first device. Here again, it helps to start with small modular units. By being small and planning to stay small, scale-up costs can be avoided and learning measured per unit of expenditure is going to be much faster than if one has to move through a progression of ever larger units.

[‡‡] $15 ton^{-1} is the low end estimate in the report, suggesting that it itself may be an attempt to get to the frictionless cost.

6 The Usefulness of Air Capture Technology

This section aims to show that the intrinsic value of air capture technology is such that the few tens of millions of dollars required to demonstrate this technology in a modular system would be well spent.

6.1 Carbon Capture from Air and Storage

Air capture can help eliminate carbon dioxide emissions related to the combustion of fossil fuels. In order to stabilize carbon dioxide concentrations in the atmosphere, essentially all the carbon dioxide that is produced from fossil fuels will have to be captured and stored. The focus of air capture will likely be on mobile sources that do not lend themselves to point source scrubbing. To the extent that air capture is used for cancelling out current or past emissions, it needs to be paired with a carbon storage technology. Storage technologies aim to keep the carbon dioxide that has been captured by any method out of the atmosphere.

From this perspective, air capture becomes just another capture technology deployed in carbon capture and storage. Air capture without a means of safe and permanent carbon dioxide disposal would be useless in managing carbon dioxide emissions related to fossil fuels. Air capture can assure that in any given year the amount of carbon dioxide captured equals or exceeds the amount emitted, which is necessary to stabilize or reduce the carbon dioxide content of the atmosphere. In a carbon neutral world, air capture may only capture a small fraction of the carbon that is stored, as point source management will usually be more economical.

A lack of safe and permanent storage that also meets with public acceptance could make it impossible to use fossil fuels in an environmentally acceptable fashion even if air capture economics proves favorable. Affordable air capture, while also relying on storage, could alleviate storage resource limitations by opening access to remote sites, where concerns over safety can be better managed than in urban areas where fossil fuels are used and carbon dioxide is emitted. Remote locations with large storage potential and readily available wind offer the possibility of collecting large amounts of carbon dioxide without inconveniencing large number of people. An example is a recent case study for the Kerguelen Islands,[39] which are located in the Southern Ocean, have an enormous wind potential and are at the center of a large igneous province (LIP) whose basalts could bind vast amounts of carbon dioxide. The wind energy available in a small section of the main island would be sufficient to dispose of 75 million tons of carbon dioxide per year.

The use of fossil fuels may eventually cease, but there may still be a residual storage demand, because it may become necessary to reduce the carbon dioxide concentration in the atmosphere below that of today. It is probably a reasonable assumption that most of the CO_2 that has left the atmosphere to end up in the ocean or the biosphere, will be gradually released again if the atmospheric partial pressure returns to earlier levels. One should therefore assume that it will take the capture of 15 gigatons of CO_2 to lower the partial pressure of CO_2 in the air by 1 ppm, rather than the 7 Gt that are equivalent to 1 ppm in the air. At present the CO_2 in the air rises by 1 ppm for every 15 Gt CO_2 emitted. While there is undoubtedly some hysteresis in the response, it would be prudent to assume that over decades the response is approximately symmetric. Thus, a rough estimate suggests that reducing CO_2 in the atmosphere by about 50 ppm, would require a 25 year effort collecting the equivalent of today's annual emissions and finding storage for about 750 Gt of CO_2. If the world would have to reduce the carbon dioxide level from 450 ppm back to 350 ppm, the size of this task would be doubled.

Air capture can be integrated into carbon capture and storage, but it will always represent the capture of last resort and the most expensive option available. If capturing carbon dioxide at a particular source would cost more than air capture, it would be advantageous to pay for the disposal of air captured carbon dioxide at a remote site instead. Conversely, if capture at the source is lower in cost, air capture is unlikely to be deployed. The advantage of air capture is that it is not limited to a particular type of carbon dioxide source and that the location of the source does not have any impact on the cost of capture or storage. However, for many sources there will be cheaper solutions and thus they are likely to deployed, even if capture from air provides a credible alternative.

6.2 Fugitive Emissions

Even above the $100 ton^{-1} threshold, the ability to remove carbon dioxide from the atmosphere can create a useful tool for eliminating fugitive emissions. Not all carbon dioxide can be captured at a source, and some of the carbon dioxide that is stored may escape by accident or through predictable leakage. Fugitive emissions can add up, and make it questionable whether fossil carbon based point sources can be allowed to operate in the future, because emissions reductions are simply not enough to approach a net zero carbon economy. For example, coal plants may only capture 90% of their emissions; they may increase consumption of coal by 25%, and incur life cycle losses in mining, preparation and transport which are typically on the order of 15% of the coal consumed.[40] With these plausible assumptions, the actual reduction in emissions compared to the baseline of the old plant is only 73%. For an individual coal plant this is less of a reduction than policy makers hope to achieve for the economy has a whole. Capture from air can cancel out fugitive emissions, albeit at higher cost than baseline scrubbing. Air capture can help point sources to approach net zero or even net negative emissions, which otherwise would not be practical.

Not only can air capture technology deal with predictable fugitive emissions, it can also deal with accidental or unplanned emissions that could result in future liabilities. By setting a price for remediation, air capture renders these losses insurable and it monetizes the risk of leakage. Monetizing the risk creates economic incentives for better technology or more careful planning.[5]

6.3 Risk Management to Oil Resource Holders

The owners of oil resources depend on the transportation sector as their primary customer. If liquid, fossil fuels were made obsolete because of climate change, the premium petroleum receives over coal would be eliminated. Based on energy content, coal is about ten times cheaper than oil. The higher price of oil can be maintained because it is easy to convert oil into gasoline, diesel or jet fuel, while it is difficult to do the same with coal. Owners of oil reserves face a unique business risk not shared by the coal and gas industries and impossible to mitigate by point source capture. Automobiles, ships and planes emit their combustion products to the air. This can only be changed by abandoning fossil fuels. On-board capture is not practical; the weight of the carbon dioxide would be too large.

Thus, the future of oil is directly intertwined with the availability of air capture. Without affordable air capture, petroleum based fuels will have to be phased out and known oil reserves in the ground would become stranded. The demonstration of direct air capture at an affordable price would provide a hedge against this risk. Affordable air capture would convert known oil deposits that exceed the carbon dioxide limit of the atmosphere from low value carbon, which at best could be burned in power plants with built-in CCS, into valuable fuels for the transportation sector. The long term ability to use oil would benefit oil field owners. More importantly, because the carbon reservoir in the air is not owned by anyone, air capture could prevent a potentially ruinous race to sell one's oil ahead of competitors who would otherwise use up the available CO_2 quota.

6.4 Managing the Risks of Global Warming

If damages inflicted by climate change were to reach crisis proportions, then cost concerns in removing carbon dioxide from the atmosphere are likely to be pushed largely aside and technologies may be introduced in an emergency that are much more expensive than what would be acceptable in normal circumstances. Nevertheless, even in an emergency the cost cannot exceed a few hundred dollars per ton, because the total cost of reduction must ultimately be affordable to society. In such a crisis air capture will compete with indirect methods of carbon dioxide removal like adding alkalinity to the ocean, or ocean fertilization. Other technologies that can stop warming directly, like solar radiation management would likely be deployed as well. However, removing carbon dioxide from the air by any means eliminates the root cause of climate change rather than simply masking some of its symptoms. Direct air capture and biomass capture have the advantage that they can deliver the carbon dioxide in concentrated form and avoid polluting other large reservoirs with the carbon residue. Unlike biomass production, air capture can conceivably reach the necessary scale without putting huge demand on available land.

However, even in an emergency, reducing the carbon dioxide content of the air is slow. At a minimum it would require decades. The more expensive the process, the longer it will take to implement. Just like it was the integral of past emissions that caused today's problems, it will be the sum of years of capture that will lead to significant reductions. It will take time to build a large fleet of collectors, and they will have to run many years before carbon dioxide can be reduced to a predetermined level at or below current levels. Air capture, even though it opens the door to a return to earlier levels of carbon dioxide in the atmosphere, is too slow and cumbersome to justify a wait-and-see attitude on climate change.

6.5 Air Capture as a Tool for Geoengineering

Direct air capture is sometimes considered a geoengineering option to maintain the temperature of the planet. It is possible that in the future air capture will be used as a planetary temperature control strategy, but as long as emissions exceed capture, air capture is not providing a thermostat on the planet but only a means of cleaning up after the end of the conventional tail pipe. Only when net emissions have been reduced to zero, can one begin to introduce temperature control strategies. Simply removing old emissions from the air still seems more clean-up than geoengineering of climate. However, if the rate of carbon dioxide removal would be driven by observed temperature changes, one would enter a realm of true geoengineering.[41]

Controlling the carbon dioxide concentration may provide a control of the earth's temperature, because greenhouse gases provide a long lever arm for manipulating energy flows on the planet. The rerouting of solar energy fluxes by carbon dioxide emissions dwarfs the energy added in the combustion process by orders of magnitude. Small inputs create large outputs. Long lever arms afford means of changing processes that otherwise would be out of human reach. However, because one lacks an even longer lever arm, there is little one could do if the climate responded in an unexpected manner. Setting in motion a dynamics is sometimes far easier than returning the system to its initial state.

6.6 Closing the Non-fossil Carbon Cycle

The atmospheric carbon budget can also be closed by recycling rather than disposing of the carbon. The advantages of liquid transportation fuels can be obtained by collecting carbon dioxide from the air, and converting carbon dioxide and water with non-fossil energy into hydrocarbon fuels. The net carbon impact on the air would be zero. Direct air capture is an enabling technology, because, unlike water, the carbon dioxide once emitted will stick with the atmosphere until it is physically removed.

The first step in producing fuel from water and carbon dioxide is to chemically reduce one or both. The removal of oxygen results in carbon monoxide and/or hydrogen and it requires an energy input which exceeds the energy content of the fuel that is eventually produced. One approach for reducing water and carbon dioxide is electrolysis.[42] High temperature electrolysis cells can reduce a mixture of steam and carbon dioxide to carbon monoxide and hydrogen.[43,44] However, there are many different approaches that have been proposed, involving different intermediate products and different sources of energy ranging from high grade heat,[45] to the direct energy in sunlight.[46]

The gas mixture of hydrogen and carbon monoxide is known as synthesis gas or "syngas". Syngas can be transformed to liquid fuels like gasoline, diesel or jet fuel *via* Fischer-Tropsch reactions, or into other chemicals like methanol, which are simpler and thus allow for the production of chemically clean products. These chemical pathways are analogous to bio-fuel production which uses green plants to extract carbon dioxide from the air, and photosynthesis to make bio-fuels. However, artificial trees are better at collecting CO_2 from the air than plants, and photovoltaic panels are far more efficient than leaves in harnessing the energy embodied in sunshine. Thus the land requirement, which would be dominated by solar energy collection would be an order of magnitude smaller than what is needed for bio-mass growth.

At least from today's perspective, this use of air capture is less sensitive to the price of air capture than using it for carbon capture and storage, mainly because the cost of electricity will overwhelm the cost of carbon dioxide capture. Assuming a 50% efficiency in converting electricity to fuel, a liter of gasoline would require 20 kWh of electricity and 2.5 kg of carbon dioxide. At $2 ¢$ kWh^{-1} and \$100 t^{-1} CO_2, the cost contributions of the two resources to a liter of gasoline would be $40 ¢$ and $25 ¢$, respectively. Even at a low electricity price and high price of air capture, the resource cost of electricity dominates the cost of carbon dioxide.

Closing the carbon cycle with air capture offers alternative liquid fuels independent of fossil carbon resources. Whether or not synthetic fuels can be economic will largely depend on the relative cost of electricity and oil. If air capture can be made to work at all, its cost will likely be small compared to the cost of the electric power input. On the other hand, the introduction of highly intermittent renewable energy would greatly benefit from a flexible electricity consumer that can absorb electric power when it is generated in excess. Crucial for such an infrastructure would be the development of low-capital-cost electrolysis systems that can afford to stand idle a large fraction of the time.

Air capture could play a major role in a world in which carbon capture and storage is important and in which air capture would be the capture of last resort. But if storage capacity proves to be limited and may have to be reserved for removing past carbon dioxide emission from the air, then the future use of fossil fuels would be in doubt, and liquid fuels would have to be abandoned or made with carbon dioxide that has been removed from the air. This would open the door to air capture based fuels and bio-fuels. Bio-fuels, ranging from corn-alcohol to advanced algae fuels, offer a means of realizing the vision of a closed carbon cycle and for that reason they have proven to be very popular, even if they still fall short on practicality. Air capture provides an alternative option to achieve the same goal. In comparison with biomass fuels, air capture technology would have a much smaller footprint in terms of land use and environmental impact, but air capture based carbon cycling would incur significant costs for the energy required to produce new fuel from carbon dioxide and water.

7 Discussion and Conclusion

Air capture is a nascent technology with large promise. It has been shown in laboratory and small prototype demonstrations that artificial trees are better at capturing carbon dioxide than natural trees. Just as a tractor out-performs a horse in pulling a plow, an artificial tree of a given size can exceed the capture rate of a natural tree by three orders of magnitude. It will not be necessary to recreate an Amazon rainforest to have impact on the anthropogenic carbon budget. Some thermal swing technologies and the moisture swing technology we have pioneered are also very energy efficient. Indeed, it would be possible, albeit not advisable, to provide the energy for air capture devices from coal fired power plants that release their emissions to the atmosphere. Roughly one third of the air capture device's capacity would then go toward re-capturing the carbon dioxide emitted in its operation.[19]

A major obstacle to the development of air capture is the lack of a compelling scenario for advancing the technology in the absence of climate change regulations. Investment of a few tens of millions of dollars would probably suffice to answer the questions the critics have raised about the viability of the technology. Technology development and demonstration at a university or government laboratory could result in publicly accessible and visible capture units that demonstrate the technology. This would make it possible to assess air capture technology and shape policy thinking.

On the other hand, air capture technology very likely will also need market drivers. A well designed price for carbon could easily accommodate air capture technology. If carbon reductions by carbon storage can be monetized in some form, air capture technology can begin to compete in an open market place. While a number of countries have carbon taxes that would be sufficient to make air capture technology economically viable, to the best of my knowledge it would require changes in these policies to allow air capture to harness the financial benefit. There is no mechanism to monetize the carbon reduction achieved with air capture and have the benefit flow back to the consumer who could pay for capturing emissions rather than pay a higher fee that permits the emission. Because the direct air capture concept is new, it is not yet integrated into current policy options.

In the meantime, air capture might be able to bootstrap itself in commercial applications in which physical carbon dioxide is of practical use. Since the introduction of air capture would have large policy implications, it might be useful to subsidize such applications, like feeding carbon dioxide to greenhouses, even though there is no immediate impact on carbon reductions or climate change.

Once air capture units have been developed, one can consider their policy implication more thoroughly. In order to get there, it will be necessary to overcome concerns of people who see air capture as a threat to their approaches to becoming carbon neutral. In some cases, air capture would indeed be a competitor. In others it may even be synergetic. From a policy perspective it is important to understand the various relationships and manage an optimal path toward climate stabilization.

Air capture will likely prove to be a formidable competitor to the hydrogen economy, because it would remove a major motivation for abandoning carbon rich fuels in favor of hydrogen. The distribution and storage cost of hydrogen far exceeds those of liquid fuels, making room for a substantial price for air capture. If hydrogen is made cheaply from coal and gasoline from petroleum, then both technologies share the need for carbon capture and storage. Liquid fuels would have to pay a premium in air capture over capture at the source, but would benefit from a much lower cost of the fuel distribution system, which in the case of hydrogen would likely involve pipelines. Pipeline costs are well understood, and even in a mature system would add a significant cost.[47] Thus, we predict that the frictionless costs of these two options likely favor air capture technology.

Artificial trees are not competing with solar, wind or nuclear energy. Indeed, they might prove extremely helpful to renewable energy systems in that they can effectively mitigate the huge intermittency in power grids dominated by renewable energy. In the case of nuclear energy the availability of air capture could simplify the power plant design by allowing the plant to operate at constant power at its optimal operating point and use excess electricity in fuel production. In both cases, air capture with fuel synthesis can become a flexible consumer of power which can help match supply to demand. Air capture with fuel synthesis could even create an electricity storage buffer in terms of the fuel produced. The roundtrip efficiency may be low, but storing electricity in batteries becomes extremely expensive if electricity is to be stored for weeks or even on a seasonal time scale. In effect one trades efficiency for capital cost, which increases linearly in the cycle time of the storage device.

Air capture is not a competitor for point source capture. The "frictionless cost" of point source capture is always lower than that of air capture and thus eventually point source capture should have a competitive edge. On the other hand, the existence of air capture could very well force power plant owners to act. With the right policies in place, air capture could set a price of carbon, and once this price becomes acceptable, owners of point sources would face the alternative of paying the air capture price or developing their own technology which once developed should be of lower cost. In the absence of a competitor, it is much easier to hide behind the likely high cost of an unproven and yet unknown technology. Air capture from this perspective has the advantage that it is not limited to a particular type of plant but can be applied to any potential point of emission. Indeed as has been suggested by Tim Fox,[48] the cost of air capture should set the price of carbon, and as the technology improves all other approaches will have to keep up with improvement that are likely to occur in air capture.

Air capture keeps the door open for liquid fuels in the transportation sector. It is a direct competitor of the fully electric car with battery storage, but air capture is not a competitor to the electric car *per se*; it may well be synergistic with it. This is best seen by comparing the frictionless cost of the all-electric car, which is charged from an outlet and carries its energy in batteries, with that of a similar hybrid car that replaces most of the batteries in the all-electric car with a small fuel cell for charging the remaining battery and a small liquid-fuel tank for storing on-board energy. By reducing weight, this would improve the performance and mileage of the car without losing any of the advantages of the electric car. The hybrid car requires far less battery and can have a far larger driving range than the all-electric car. The hybrid also uses far less exotic resources and ultimately should be cheaper and thus more likely competitive with an internal combustion engine than the all-electric car.

The hydrocarbon fuels used in such vehicles need not be petroleum based. They could be produced from other fossil fuels, or made synthetically from non-fossil energy sources and air-captured carbon dioxide. Air capture in this broad sense could include biomass based air capture combined with bio-fuel production. However, it is worth pointing out that direct air capture tends to be much more efficient than biomass based capture. A natural tree collects a fraction of a percent of the carbon dioxide that an artificial tree of similar frontal area would collect.

Whether fossil fuels can continue to be used may very well depend on the perceived storage capacity for carbon dioxide. If this storage capacity seems limited, then the little that is available may have to be reserved for several hundred or even thousand gigatons of carbon dioxide that the world will likely have to recover in order to restore the climate to a safe level. Even without fossil carbon and associated carbon storage, air capture would be necessary as long as carbonaceous fuels are used. The advantage of liquid carbonaceous fuels for energy storage and energy transport is so large that it is worth paying a significant premium for these fuels. This has already been demonstrated in the market, as oil is ten times as expensive as coal, and it likely would remain true in a world that has large amounts of very low cost, but intermittent electricity resources in terms of wind or solar energy. Air captured fuels could provide the energy of the transportation sector and also provide the energy source for a backup system that is needed when the sun is not shining and the wind is not blowing.

There is generally a large concern that the availability of air capture creates a moral hazard, because its ability to reverse emissions makes it easier to postpone action. In my view this debate is over, now that the IPCC has noted the need for negative emissions in various scenarios. The new IPCC report states clearly, "A large fraction of climate change is largely irreversible on human time scales, unless net anthropogenic carbon dioxide emissions were strongly negative over a sustained period."[49] All negative emissions technologies would share this moral hazard and air capture is a central and enabling technology for negative emissions. It will not act fast, but deployed on the scale of current consumption it could produce a measurable difference in a few decades. Negative emissions do not provide an excuse for delay. Instead, delays already encountered make negative emission technologies necessary.

The availability of reversing emissions can fundamentally change the international negotiations on climate change. The major ingredient in a tragedy of the commons is that cooperation of all is a necessity. For example, nuclear proliferation threats cannot be eliminated unless all nations cooperate. The climate change dilemma has been seen in the same light. As long as emissions are considered irreversible, it is insufficient for a large number of countries to agree that emissions should be reduced. Free-riders will ruin the climate more slowly but just as completely. However, air capture technology makes it possible to recapture emissions. Therefore, the emissions of free-riders can be reversed at a cost that can be quantified. This creates a quantifiable grievance and a potential demand for compensation in trade negotiations that could be avoided at lower cost by the free-rider's cooperation. Air capture would give cooperating countries assurance that the problem can be solved even in the presence of a few free-riders, who ultimately could be confronted with a bill for cleaning up.

Air capture, if demonstrated to be viable and cost-effective, would not only provide a tool for managing carbon, but its very presence and nearly universal applicability would create an environment in which the introduction of more focused and often more affordable carbon management technologies could thrive.

Air capture will need to be considered as part of a larger system, where it is integrated into a wider energy infrastructure. It has a role in removing excess carbon from the environment, for which it needs to be paired with safe and permanent disposal options. Air capture could also contribute to balancing out day-to-day emissions of fossil carbon dioxide where it again needs to be paired with a disposal option with an even larger storage capacity. Lastly air capture can enable the continued use of liquid carbon based fuels for which it needs to be paired with technologies to transfer non-fossil energy into chemicals produced from carbon dioxide and water. One should not think of air capture as a stand-alone technology but as part of a larger infrastructure system that still needs to be developed. Its presence will make for a richer set of options and opportunities and a smoother path from today's energy systems to future systems, as the necessary additions can be isolated and user interfaces to the energy infrastructure can be left largely intact.

Acknowledgements

The author gratefully acknowledges financial support from Gerry Lenfest for research on air capture. The author is also a co-founder of Kilimanjaro Energy Inc., a company located in San Francisco that develops air capture technology. The author advises the company and owns shares in the company.

References

1. S. Solomon, G.-K Plattner, R. Knutti and P. Friedlingstein, Irreversible climate change due to carbon dioxide emissions, *Proc. Natl. Acad. Sci. U. S. A.*, 2009, **106**(6), 1704–1709.
2. K. S. Lackner, H.-J. Ziock and P. Grimes, Carbon dioxide extraction from air: is it an option?, in *Proceedings of the 24th International Conference on Coal Utilization & Fuel Systems*, Clearwater, Florida, 1999.
3. D. Keith, M. Ha-Duong and J. Stolaroff, Climate srategy with CO2 capture from the air, *Clim. Change*, 2006, **74**(1), 17–45.
4. D. Keith, Why capture CO2 from the atmosphere?, *Science*, 2009, **325**(5948), 1654.
5. K. Lackner and S. Brennan, Envisioning carbon capture and storage: expanded possibilities due to air capture, leakage insurance, and C-14 monitoring, *Clim. Change*, 2009, **96**(3), 357–378.
6. C. Azar, K. Lindgren, M. Obersteiner, K. Riahi, D. P. van Vuuren, K. M. G. den Elzen, K. Möllersten and E. D. Larson, The feasibility of low CO2 concentration targets and the role of bio-energy with carbon capture and storage (BECCS), *Clim. Change*, 2010, **100**(1), 195–202.
7. J. H. Martin, Glacial-interglacial CO2 change: the iron hypothesis, *Paleoceanography*, 1990, **5**(1), 1–13.
8. K. O. Buesseler and P. W. Boyd, CLIMATE CHANGE: will ocean fertilization work?, *Science*, 2003, **300**(5616), 67–68.
9. K. Caldeira and G. H. Rau, Accelerating carbonate dissolution to sequester carbon dioxide in the ocean: geochemical implications, *Geophys. Res. Lett.*, 2000, **27**(2), 225–228.
10. K. S. Lackner, Carbonate chemistry for sequestering fossil carbon, *Annu. Rev. Energy Environ*, 2002, **27**(1), 193–232.
11. K. S. Lackner, Washing carbon out of the air, *Sci. Am.*, 2010, **302**(6), 66–71.
12. G. H. Rau, CO2 mitigation via capture and chemical conversion in seawater, *Environ. Sci. Technol.*, 2010, **45**(3), 1088–1092.
13. K. S. Lackner, S. Brennan, J. M. Matter, A.-H. A. Park, A. Wright and B. van der Zwaan, The urgency of the development of CO2 capture from ambient air, *Proc. Natl. Acad. Sci. U. S. A.*, 2012, **109**(33), 13156–13162.

14. R. Socolow, M. Desmond, R. Aines, J. Blackstock, O. Bolland, T. Kaarsberg, N. Lewis, A. P. Marco Mazzotti, K. Sawyer, J. Siirola, B. Smit and J. Wilcox, *Direct Air Capture of CO2 with Chemicals. A Technology Assessment for the APS Panel on Public Affairs,* American Physiccal Society, Washington DC, 2011.
15. D. Sarewitz and R. Nelson, Three rules for technological fixes, *Nature,* 2008, **456**(7224), 871–872.
16. T. K. Sherwood, Mass transfer between phases, in *33rd Annual Priestley Lecture, sponsored by Phi Lambda Upsilon and Associated Departments,* Pennsylvania State University, 1959, 86.
17. K. House, A. Baclig, M. Ranjan, E. van Nierop, J. Wilcox and H. Herzog, Economic and energetic analysis of capturing CO_2 from ambient air, *Proc. Natl. Acad. Sci. U. S. A.,* 2011, **108**(51), 20428–20433.
18. C. J. King, *Separation & Purification: Critical Needs and Opportunities,* National Academy Press, 1987.
19. K. S. Lackner, Capture of carbon dioxide from ambient air, *Eur. Phy. J., Special Topics,* 2009, **176**, 93–106.
20. S. Choi, J. H. Drese, P. M. Eisenberger and C. W. Jones, Application of amine-tethered solid sorbents for direct CO_2 capture from the ambient air, *Environ. Sci. Technol.,* 2011, **45**(6), 2420–2427.
21. J. A. Wurzbacher, C. Gebald and A. Steinfeld, Separation of CO_2 from air by temperature-vacuum swing adsorption using diamine-functionalized silica gel, *Energy Environ. Sci.,* 2011, **4**, 3584–3592.
22. M. D. Eisaman, L. Alvarado, D. Larner, P. Wang, B. Garg and K. A. Littau, CO_2 separation using bipolar membrane electrodialysis, *Energy Environ. Sci.,* 2011, **4**, 1319–1328.
23. M. Mahmoudkhani and D. W. Keith, Low-energy sodium hydroxide recovery for CO_2 capture from atmospheric air? Thermodynamic analysis, *Int. J. Greenhouse Gas Control,* 2009, **3**(4), 376–384.
24. K. S. Lackner, The thermodynamics of direct air capture of carbon dioxide, *Energy,* 2013, **50**, 38–46.
25. R. V. Davies, J. Kennedy, R. W. McIlroy, R. Spence and K. M. Hill, Extraction of uranium from sea water, *Nature,* 1964, **203**(4950), 1110–1115.
26. F. Best and M. Driscoll, Prospects for the recovery of uranium from seawater, *Nucl. Technol.,* 1986, **73**(1), 55–68.
27. E. Schneider and D. Sachde, The cost of recovering uranium from seawater by a braided polymer adsorbent system, *Sci. Global Security,* 2013, **21**(2), 134–163.
28. T. Wang, K. S. Lackner and A. Wright, Moisture-swing sorption for carbon dioxide capture from ambient air: a thermodynamic analysis, *Phys. Chem. Chem. Phys.,* 2013, **15**, 504–514.
29. T. Wang, K. S. Lackner and A. Wright, Moisture swing sorbent for carbon dioxide capture from ambient air, *Environ. Sci. Technol.,* 2011, **45**(15), 6670–6675.

30. W. Li, S. Choi, J. H. Drese, M. Hornbostel, G. Krishnan, P. M. Eisenberger and C. W. Jones, Steam-stripping for regeneration of supported amine-based CO2 adsorbents, *ChemSusChem*, 2010, **3**(8), 899–903.
31. E. Dahlgren, C. Göçmen, K. S. Lackner and G. van Ryzin, Small modular infrastructure, *Eng. Economist, in press*.
32. K. S. Lackner, S. Brennan, J. M. Matter, A.-H. Park, A. Wright and B. Van Der Zwaan, The urgency of the development of CO2 capture from ambient air: supporting information, *Proc. Natl. Acad. Sci. U. S. A.*, 2012, **109**(33), 13156–13162.
33. K. S. Lackner, P. Grimes and H. J. Ziock, *Carbon Dioxide Extraction from Air?*, Los Alamos National Laboratory, Los Alamos, NM, 1999.
34. International Organization of Motor Vehicle Manufacturers, oica.net. (accessed 10 November 2013).
35. W. D. Nordhaus, Do real-output and real-wage measures capture reality? The history of lighting suggests not, in *The economics of new goods*, University of Chicago Press, 1996, 27–70.
36. H. Hammar and M. Sjöström, Accounting for behavioral effects of increases in the carbon dioxide (CO2) tax in revenue estimation in Sweden, *Energy Policy*, 2011, **39**(10), 6672–6676.
37. R. Baciocchi, G. Storti and M. Mazzotti, Process design and energy requirements for the capture of carbon dioxide from air, *Chem. Eng.Process*, 2006, **45**(12), 1047–1058.
38. B. Metz, O. Davidson, H. D. Coninck, M. Loos and L. Meyer, *IPCC Special Report on Carbon Dioxide Capture and Storage*, IPCC, Cambridge University Press, New York, 2005.
39. D. S. Goldberg, K. S. Lackner, P. Han, A. L. Slagle and T. Wang, Co-Location of air capture, subseafloor CO2 sequestration, and energy production on the Kerguelen Plateau, *Environ. Sci. Technol.*, 2013, **47**(13), 7521–7529.
40. N. A. Odeh and T. T. Cockerill, Life cycle GHG assessment of fossil fuel power plants with carbon capture and storage, *Energy Policy*, 2008, **36**(1), 367–380.
41. P. Eisenberger, Closing the Global Carbon Cycle: Climate Stabilization and Sustainable Carbon Energy, in preparation.
42. C. Graves, S. D. Ebbesen, M. Mogensen and K. S. Lackner, Sustainable hydrocarbon fuels by recycling CO2 and H2O with renewable or nuclear energy, *Renew. Sustain. Energy Rev.*, 2011, **15**(1), 1–23.
43. S. H. Jensen, P. H. Larsen and M. Mogensen, Hydrogen and synthetic fuel production from renewable energy sources, *Int. J. Hydrogen Energy*, 2007, **32**(15), 3253–3257.
44. C. Graves, S. D. Ebbesen and M. Mogensen, Co-electrolysis of CO2 and H2O in solid oxide cells: performance and durability, *Solid State Ionics*, 2011, **192**(1), 398–403.
45. J. Kim, T. A. Johnson, J. E. Miller, E. B. Stechel and C. T. Maravelias, Fuel production from CO2 using solar-thermal energy: system level analysis, *Energy Environ. Sci.*, 2012, **5**, 8417–8429.

46. N. S. Lewis and D. G. Nocera, Powering the planet: chemical challenges in solar energy utilization, *Proc. Natl. Acad. Sci. U. S. A.*, 2006, **103**(43), 15729–15735.

47. J. M. Ogden, Prospects for building a hydrogen energy infrastructure, *Annu. Rev. Energy Environ.*, 1999, **24**(1), 227–279.

48. T. A. Fox, Energy innovation and avoiding policy complexity: the air capture approach, *Energy Environ.*, 2012, **23**(6), 1075–1092.

49. M. Collins and R. Knutti, Working Group I contribution to IPCC Fifth Assessment Report on Climate Chante, in *IPCC Fifths Assessment Report*, IPCC, 2013.

Cooling the Earth with Crops

TARAKA DAVIES-BARNARD

ABSTRACT

Food security and climate change are two of the biggest challenges which face humanity in the 21st Century and agricultural land is the physical interface for these interlinked issues. This chapter addresses how cropland interacts with climate; the ways in which crops have affected climate in the past; and how crops could help mitigate climate change in the future. Of the ways that climate issues and crops are related, one of the most relevant to the future is through geoengineering. The concept of deliberately using crops to reduce the surface air temperature is still in development, but has gathered considerable interest in recent years. Models suggest that in North America and Europe, a moderate increase in crop albedo could decrease summertime temperatures by up to 1 °C. Although this amounts to a small change compared with many other geoengineering proposals, it could be made with relatively little cost and would make a significant difference to crops which are particularly sensitive to high temperatures, such as wheat. Along with other climate mitigation strategies, cooling with crops could be one aspect of a deliberate policy to limit the dangerous impacts of climate change.

1 Introduction

Agricultural land currently covers 37% of the world's land surface,[1] and most projections indicate that there will be future increases.[2] Since crops represent a significant proportion of this anthropogenically altered land cover, they have substantial potential as a platform for land surface based climate

Issues in Environmental Science and Technology, 38
Geoengineering of the Climate System
Edited by R.E. Hester and R.M. Harrison
© The Royal Society of Chemistry 2014
Published by the Royal Society of Chemistry, www.rsc.org

solutions. Cropland has multiple ways in which it could be used to help
mitigate climate warming, including conventional mitigation (reducing
carbon emissions from the agricultural sector),[3] as a source for carbon
capture and storage (CCS) geoengineering[4] (for instance using bio-char)[5]
and as a way of managing surface net solar radiation (surface SRM geoengineering)[6] by altering the albedo of cropland. Any of these crop based ideas
could be considered as 'cooling with crops'. However, the most commonly
discussed idea as a geoengineering method using crops is bio-geoengineering, which proposes higher leaf albedo crops as a way of creating lo-
calized cooling.[7]

Bio-geoengineering leaf albedo increase would be a small part of much
larger anthropogenic changes to surface albedo, and other surface properties, which have occurred over more than 1000 years.[8] Forest clearance for
agriculture and the subsequent intensification of agriculture mean that
crops have affected the regional and global climate substantially, *via* their
alteration of the land surface. Similarly, future changes to land use and their
consequent changes to land cover will have significant affects on the climate.
These changes affect climate inadvertently but may be equal or larger in
magnitude to the projections of deliberate interventions such as bio-
geoengineering.

In this chapter the mechanisms by which crops can cool the Earth are
reviewed, focusing on past and projected future changes to climate from
crops and then looking at how future cooling with crops could be achieved
and what the implications might be.

2 Mechanisms

The ways which surface properties affect climate can be categorised as
biogeophysical or biogeochemical. Biogeochemical properties of the land
surface are typified by changes to atmospheric composition from the
emissions of greenhouse gases (*e.g.* carbon dioxide, methane, nitrous oxide)
from the land surface. Biogeochemical land surface changes impact the
climate by changing the atmospheric greenhouse gas composition, which
affects the amount of outgoing longwave radiation, changing the energy
balance. The land surface is currently a net carbon sink, absorbing around
2.4 gigatonnes of carbon per year.[9] Changes in land cover, especially deforestation, could alter the size of this sink. Similarly, warming could open
up carbon stores such as methane from thawing of permafrost.[10] Russian
permafrost regions alone contain 50 gigatonnes of carbon and mid century
could account for a 0.012 °C global temperature rise. Changes in emissions
of aerosols can also be considered biogeochemical land surface changes.

The biogeophysical properties can be understood as the physical changes
to the land surface which affect the energy and momentum balance directly,
rather than through changes in the atmospheric composition. The net radiative fluxes are made up of the net short wave and longwave radiation (see

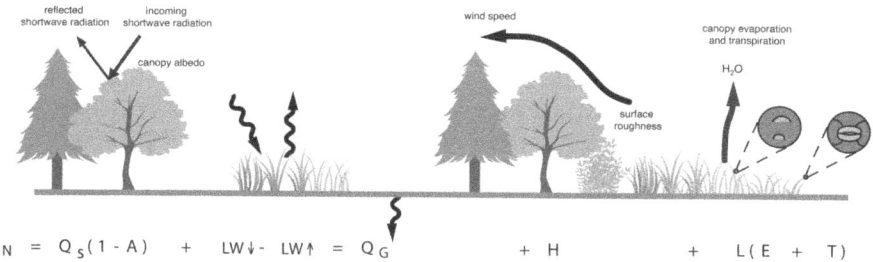

$$R_N \; = \; Q_S(1 - A) \quad + \quad LW\!\downarrow - \; LW\!\uparrow \; = \; Q_G \qquad\qquad + \; H \qquad\qquad + \quad L\,(\,E \; + \; T\,)$$

Figure 1   Representation of the biogeophysical parameters in the surface energy balance which are affected by changes in land surface cover. The net radiative fluxes (R_N) are determined by the net short and longwave radiation fluxes. The albedo affects the amount of incoming shortwave solar radiation which is reflected back out of the atmosphere (1–A). The surface emissivity is the amount of longwave radiation which is emitted back from the surface (LW $\uparrow$). These two make up the direct removals from the energy budget from the total incoming solar radiation. On the right hand side of the equation, the surface roughness affects the energy balance through the sensible heat flux (H), some energy goes into heating the soil (Q_G) and the latent heat flux (L) is made up of evaporation (E) from the canopy and soil, and transpiration (T) from plant photosynthesis.

Figure 1). These net radiative fluxes are partitioned at the surface into: heat flux into the soil; latent heat (as either evaporation or transpiration); or sensible heat (heat exchange due to the effect of differing temperatures) which may remain or be moved into other areas as convective heat (see Figure 1.) The balance of which of these factors is most important to the resultant temperature varies with latitude and season, as well as the individual surface properties themselves.[11]

2.1   Biogeophysical Mechanisms

2.1.1   Albedo.   Albedo is a measure of the ratio of radiation reflected from a surface to the total amount of radiation incident upon it. The exact albedo is dependent on the amount of incoming solar radiation, making it very difficult to calculate. A range of other measurements and terms are used to represent albedo. For most practical applications, albedo is calculated as the bidirectional reflectance distribution function over a particular range of wavelengths, as opposed to field albedo, which is the value for the entire spectrum of the solar radiation.

Changes to surface albedo are some of the largest in the surface energy budget. This is because albedo is the key parameter in the energy balance which determines the net radiative flux. It also has a large range of values, with land surface albedos varying significantly. For example, snow covered surfaces have very high albedos of about 0.9 and reflect most of the incoming shortwave radiation, whereas water covered surfaces (such as inland lakes) scatter or absorb most of the radiation and so generally have low albedos of less than 0.1. Vegetated surfaces have relatively low albedos, with grasses around 0.19–0.27 and coniferous forest lower at 0.11–0.14.[12] Therefore an increase in the surface albedo *via* a change from trees to grasses gives an albedo cooling effect. The size of the albedo feedback can be large, especially for changes between snow covered and non-snow covered surfaces.

2.1.2 Evapotranspiration. The amount of evapotranspiration from the surface affects the latent heat flux (see Figure 1). Evapotranspiration is an amalgamation of two water processes, which work at very different scales temporally and have different sources.[13] Transpiration is a delayed feedback of precipitation to the atmosphere, as the water must infiltrate into the soil, be absorbed by the plant roots and transported up the plant to be transpired. Evaporation from the vegetation canopy provides a much quicker feedback because a large quantity of water can be caught in the canopy and is readily available for evaporation. Soil moisture can also evaporate quickly compared to transpiration, though because of infiltration and lower wind speeds at the surface, is closer in timescale to transpiration. Since trees have more biomass to support than grasses or crops and have deeper roots, theoretically they transpire more water. The larger leaf area of trees can often hold more water in the canopy than grasses. Therefore replacing trees with crops is usually associated with a decrease in evapotranspiration, giving a warming effect because less energy transfers to latent heat are made, and thus the energy goes into sensible heat.

2.1.3 Emissivity. The emissivity of a surface determines how much longwave radiation it will emit (see Figure 1), from the shortwave radiation the ground absorbs during the day. The importance of emissivity is most obvious at night when longwave radiation dominates the radiative budget. The higher the emissivity of a surface, the more absorbed energy it emits. A perfect black body (emissivity of one) emits energy at the theoretical rate given by the Stefan–Boltzmann equation. The range of emissivity values is small. For instance, desert soils have a relatively low emissivity, of approximately 0.9 whereas vegetated surfaces have higher emissivities of between 0.96–0.98.[14] Emissivity of a vegetated surface is related to the vegetation density and structure. The emissivity generally increases with leaf area index, giving higher emissivity for trees, on average, than grasses or crops.[15] This implies a slight warming effect when crops replace trees. The emissivity effect is relatively small and other atmospheric factors, such as cloud cover, are often much more important for the total outgoing longwave radiation.

2.1.4 The Aerodynamic Roughness. The aerodynamic roughness length describes the height at which the wind speed theoretically becomes zero. For vegetation, it is essentially related to the canopy height, but is not simply the height of the vegetation.[16] Typical values of roughness length may be 0.0002 m for open water, 0.03 m for grass, 0.1–0.25 m for crops, and 1.0 m for a forest. Lower values of roughness length allow increased wind speed across a surface, affecting the turbulent flow in the boundary layer atmosphere. This results in more convective heat transfer, and thus a cooler surface temperature as the sensible heat or evaporative cooling is enhanced.

3 Geographical Differences

Changes in land surface and vegetation can have spatially varying effects; high and low latitudes have different responses. Both temperature and precipitation can be strongly affected in different ways in different regions from changes to land cover. Crucial to the geographical variation in changes from land use is that whereas land carbon emissions from a particular location are quickly well mixed into the global atmosphere, biogeophysical effects are predominantly local. Therefore, even when the carbon emissions in a region may give a larger global change in temperature, the regional signal can still be dominated by albedo or other biogeophysical effects. The differences in impact of biogeophysical land use change are here categorized as low latitude (tropical) and high latitude (temperate and boreal), which provide contrasting responses to biogeophysical forcing.

3.1 Tropics

In the low latitudes, changes in the land surface from trees to crops generally give a warming signal.[11] A key component in this is the effect of changes in evapotranspiration. The water's change of state requires energy, which then reduces the amount of energy available for sensible heat at the surface. Therefore climate models simulating deforestation in the tropics, particularly in the Amazon, give significant warming in that region.[17] Evapotranspiration is an especially strong feedback in the tropics and has far reaching consequences for regional and global climate.[18,17] The consequence of loss of evapotranspirative cooling because of tropical deforestation is consistently found to be larger than any potential gains from increased albedo or reduced surface roughness.

Changes in tropical forest cover can have positive feedbacks at both the regional and global scale. Deforestation could potentially tip the region into a different bioclimatic regime. For instance, the increased water deficit created by Amazon tropical forest deforestation, especially when combined with deforestation of Cerrado (tropical savanna), can create a positive feedback of forest dieback.[19] The deforestation changes the local climate, which may then be combined with large scale changes which affect global climate change, making the climate of the whole region drier and leading to further forest dieback where forest remains. This in turn creates further changes to the regional climate, altering the whole bioclimatic regime in the region. This is important because the changes in water and energy fluxes in the Amazon have teleconnections worldwide which amplify the warming feedbacks.[20] For instance, changes in the deep tropical atmospheric convection in the Amazon lead to changes in storm-tracks and a northward shift in the Ferrel cell (a large scale atmospheric circulation feature) which causes warming in Europe.[20] This makes local biogeophysical changes to tropical forests very important not just for the regional climate and biosphere, but also for the world climate.

This evapotranspirative cooling acts both directly on the surface energy balance as well as affecting cloud formation in the tropics and seasonal convective rainfall in the mid latitudes.[21] Trees in the tropics have a large role in maintaining continental precipitation. For instance, in the Amazon, up to 50% of precipitation is sustained by water recycling through vegetation.[22] Water drawn from the soil water, sometimes as much as 8 meters deep, is transpired by the vegetation and returns into the atmosphere.[23] Therefore, forests, with high leaf area indexes, deeper rooting depths and greater water requirements than grasses or crops have greater transpiration and potential for water recycling. As well as transpiration, the larger canopy water capacity of trees results in more water intercepted and evaporated from the canopy. This can lead to increased cloud formation because of increased levels of water vapour in the atmosphere, consequently leading to increased rainfall. This process provides quick recycling of precipitation back to the atmosphere, which helps maintain continental rainfall.[13]

These biogeophysical warming and precipitation effects are in addition to the strong carbon feedbacks from deforestation in the tropics. It is estimated that 55% of the worlds terrestrial carbon is stored in tropical forests and are therefore an important store and sink of carbon.[9] Deforestation and forest fires can release this stored carbon,[24] both of which are projected to increase under future climate change.[25,24] The combined biogeochemical and biogeophysical effects of tropical deforestation make crop growing in the Amazon and other tropical forest areas a 'no-win' scenario.[26]

The strong impact of changes in evapotranspiration in the tropics, combined with the potential for albedo increase to feedback into lower evapotranspiration because of reduced energy at the surface,[27] mean that increased crop albedo may not be a good policy for the tropics. Deforestation for cropland would be particularly damaging. Conversely, policies which avoided warming by preventing Amazon deforestation would be a deliberate and significant contribution to cooling the planet.

3.2   Temperate and Boreal

At higher latitudes, deforestation tends to give a net cooling, because of the strong impact of the albedo changes. For boreal forest especially, a change between forest and grass or cropland has a cooling effect because of the different ways that snow lies on trees and crops.[28] Snow has a very high albedo (around 0.9) and when it completely covers a surface (*i.e.* grass or bare soil as opposed to trees), it increases the albedo considerably. For surfaces which are rougher, and therefore snow can fall through the canopy, the albedo is increased only slightly. Therefore the 'snow covered' albedo of trees as opposed to crops is very different.[29] Extensive deforestation increases summertime albedo by just 2% and wintertime in the presence of snow by more than 10%.[30] This means that the change between high latitude albedo of trees and crops is larger than the physiological albedo differences between the two plant types. The effect of this albedo change may potentially be big enough to offset the reduced carbon sink from boreal deforestation or the increased carbon sink from Boreal afforestation.[28]

Even where the albedo effect isn't enhanced by the presence of snow, it is still a crucial feedback in the mid to high latitudes. Whereas in the tropics the evapotranspirative effect and carbon emissions from deforestation are particularly strong, the lower temperatures in the mid and high latitudes mean that the cooling effect of evapotranspiration less important. Since the mid latitudes have a relatively strong response to albedo changes, this makes increased crop albedo a viable proposition. Temperature changes from trees to crops are likely to be larger for areas of substantial snow cover (*i.e.* in areas further north). Regions regularly covered with snow may not be areas where reduced temperatures would aid crop yields but might help create a critical amount of cooling in major crop growing areas or even seasonal sea ice.

4 Historical Land Cover Change

Around 10 000 years ago the Neolithic revolution began the move from predominantly forested land to the cropland we have today as shown in Figure 5(a).[1] The earliest estimates of anthropogenic climate changes from crops are put forward by Ruddiman's early anthropocene theory. Changes in climate from around 7000 years ago may be attributed to methane, emitted from growing rice, and carbon emissions from forest clearance.[31] These changes in both climate and in greenhouse gas emissions are extrapolated from proxies, so there are considerable uncertainties associated with them. Over the last 1000 years, the estimates are more reliable and an estimated deforestation of 18 million km^2 (about 12% of the land surface) has occurred. Climate model simulations suggest that this has decreased the global mean annual temperature by between -0.25 to -0.13 °C.[32] This cooling mainly originates from the last 200 years, when there was extensive deforestation. Models show that the effect on temperature of this land use change scales approximately linearly with removal of tree cover.[33] However, the mechanisms which give this result are from a range of conflicting climate signals of similar magnitude.

The historical change in land use over the last 150 years is estimated to result in a radiative forcing from albedo change of -0.2 Watts m^{-2}, (± 0.2 Watts m^{-2}), whereas changes in carbon emissions from land use change in the same period gave a radiative forcing of $+0.55$ Watts m^{-2}

($\pm$ 0.17 Watts m^{-2}).[34] Generally speaking, the historical change from trees to crops has a cooling effect *via* the biogeophysical mechanisms and a warming effect from the biogeochemical. This is because the majority of historical deforestation has been in the northern hemisphere mid-latitudes and therefore the cooling effect of the increased albedo when trees are converted to grass or cropland, predominates. However, it is still unclear whether the biogeophysical or biogeochemical effect dominates and determines the net impact to the climate. Two studies using coupled climate models find a net increase of around +0.15 °C from land use change over the last 150 years,[35,36] but a study using lower resolution earth system models of intermediate complexity found a small net decrease in temperature of −0.05 °C.[37] The estimations of biogeophysical cooling which give these results vary considerably, ranging from just −0.03 °C,[35] to −0.26 °C.[37] Although the coupled climate models agree closely on the net signal, the individual signals are more uncertain. This suggests that the biogeophysical changes to land cover can have a significant impact on climate but the size of that impact historically, and whether it is partly or wholly mitigated by the carbon emissions in the same period, is still debatable.

As well as the spatial differences in the impacts of land use change, there are also temporal differences, which make extrapolating the longer term trend more challenging. The latter part of the 20th Century saw a slight reversal of the cooling trend from biogeophysical land use change because of mid latitude afforestation.[32] The afforestation gave the opposite effect to deforestation, with a cooling signal from the biogeochemical reduction of carbon emissions and a warming from decreased albedo from trees rather than grasses. This period also saw an acceleration of tropical land use changes with large amounts of deforestation and substantial carbon emissions. Although the pace of Amazon deforestation slowed for five years in the 2000s,[38] deforestation rates are again rising and remain a serious issue.[39,40] The Amazon deforestation alone has given a detectable warming signal.[41] More recent land cover change does not necessarily follow the same pattern as previous historical changes, but can be easier to attribute due to satellite and other global data sources.

Some part of the uncertainty about the effect of past land use change is from differences in estimates of the land cover itself, both past and present. Different sources of data can result in a considerable range of possible land cover changes since 1765.[42] Even recent past (2001–2005) crop and pasture land cover estimates can differ by over 100% regionally and these differences can result in up to 0.21 °C differences in the mean annual global temperature and as much as 5° C locally.[42,43] This means that the estimates of land use change driven temperature change are uncertain. There are many idealized simulations of natural vegetation which can be used to estimate anthropogenic impact, but this approach comes with its own assumptions and uncertainties.

Future changes to the land surface, whether deliberate or as a side effect of other changes, must be seen in the context of these past changes. Although there is uncertainty about the exact scale of the biogeophysical changes from land use change in the past, the range of estimates suggests that this has been an important factor in determining present climate, especially regionally in the mid latitudes. Since most historical land use changes occurred in Europe and North America, they gave a stronger albedo feedback relative to an equivalent low latitude change but have less carbon changes associated. Future deforestation is likely to be more in tropical areas, which has very different regional and global impacts. This gives new challenges, but can still be usefully informed by historical analysis which gives insight into the spatial, temporal and data uncertainties.

5   Future Land Cover Change

Future land use change is likely to be determined by factors which influenced the temporal and spatial patterning of historical land cover change, as well as new factors relating to climate change and climate change mitigation. Agricultural productivity, population growth and trade will continue to be important. New factors such as carbon emissions targets through land carbon valuation, biofuels and carbon sequestration will also likely affect the land cover. All of the factors affecting land use change are essential to economic projections and thus future land use change scenarios are often associated with economic projections. In turn, these projections are associated not just with future land use change scenarios but are used to create climate change scenarios. The two sets of scenarios used by the IPCC (Intergovernmental Panel on Climate Change) in the *IPCC 4th and 5th Assessment Reports* give a range of possible land use futures which show some of the issues surrounding changes to the land surface and their affect on climate.

The scenarios presented in the *Special Report on Emissions Scenarios (SRES)* are story-line projections which envisage worlds with different futures,[44] which were used for the third Climate Model Intercomparison Project (CMIP) and the *4th IPCC Assessment* Report.[34] These projections are based on146different scenarios of technological and social changes, rather than achieving a particular climate outcome. The land cover changes in these scenarios significantly alter the regional climate.[45] The high carbon dioxide A2 scenario has substantial agricultural expansion which cools the mean global climate by two degrees but gives a net warming in the Amazon, as suggested by other studies referred to in section 3.1. In contrast, land abandonment in the B1 low carbon dioxide scenario gives a 1 °C warming from the land use change.[45]

The Representative Concentration Pathways (RCPs) used for *CMIP5* and the *IPCC 5th Assessment Report* also have vastly different land surface changes.[46] The RCPs use integrated assessment models to model the socio-economic paths which achieve certain climate outcomes. Unlike the scenario driven *SRES* storylines, the RCPs have explicit inclusion of climate mitigation policies where they are required to achieve the particular climate forcing aim. There are four RCP scenarios: RCP2.6, RCP4.5, RCP6.0 and

RCP8.5.[47] The numbers refer to the radiative forcing in Watts m^{-2} at the end of the century. The RCPs projections vary from small decreases in cropland to substantial increases.[48] Statistically significant differences in the climate with and without the assumed land use changes in RCP2.6 and RCP8.5 exist at a regional, though not the global, scale. Results from earth system models differ, but the net regional mean annual effect of land use change is as much as −0.2 °C in RCP8.5 from 2070–2100.[49] However, the change in land cover is not consistent with the change in total radiative forcing. The highest and lowest levels of radiative forcing (RCP2.6 and RCP8.5) are associated with high levels of deforestation in order to grow crops. By comparison, RCP4.5 and RCP6.0 both have much lower levels of deforestation and even some afforestation. This non-linearity is caused by three key aspects of the assumptions in the scenarios created by the integrated assessment models: yield increases; biofuel use; and land carbon valuation and population.

The integrated assessment models assume year on year increases in crop yield (also known as agricultural productivity growth). Most of the scenarios take their yield increases from the Food and Agricultural Organization of the United Nations estimates until 2035, which projects around a quarter of a percent increase globally in crop yield per year.[48,50] However, estimates of future yields under climate change are highly uncertain, with little clarity on even whether they are likely to be negative or positive.[51] The level of crop yield determines to a great extent how much cropland is needed. With populations peaking at between 9–12 billion in the RCPs,[52–54] without substantial increases in agricultural productivity, considerably more cropland is needed to meet demand. Conversely, lower population projections with higher crop yield increases require less land to be converted to cropland. The biogeophysical aspect of different levels of yield increases can make a significant impact on climate. For instance, the RCP4.5 scenario with no yield increases gives a mean annual climate −0.37 °C cooler than no land use change in North America. In a no mitigation strategy scenario, similar to RCP8.5, the mean annual cooling is up to −1.62 °C regionally.[55] The changes in land carbon emissions cancel out these effects globally, but residual regional effects would be likely to remain. Due to the regional differences in land use change, the effect of low yield increases or yield decreases will depend on where the cropland expansion takes place.

Biofuel use also results in increased competition for land and therefore pressure to deforest, as it increases the total amount of cropland needed. Biofuel use varies in the RCPs, but is an important element of the mitigation.[54] There are some synergies between carbon emissions targets and the biogeophysical impacts of land use change because of the avoided fossil fuel emissions and increased albedo in the mid latitudes. However, at low latitudes the carbon savings from biofuels may be less than the impact of the deforestation, giving a net warming.[56] Further, there are differences in the way that biofuels are specified in the models (as trees or grass crops). In RCP8.5, biofuels are categorized as wood, whereas the other three scenarios categorize them as crops.[48] These categorizations will be important for the biogeophysical effects of land use change and therefore are terms that need to be clarified.

As well as being affected by demand for cropland itself, cropland extent is also affected by demand for competing land covers, notably forest. RCP4.5 includes the carbon emissions from the land in the total accountable carbon emissions, making afforestation a feasible mitigation strategy. This creates a counter-incentive to cropland increases required by population increases. Though the amount of afforestation in this universal carbon tax scenario is modest, the avoided deforestation is large when compared to a fossil fuel only carbon tax.[53,57] By comparison, the other RCPs do not account for the carbon emitted from land cover changes, and therefore are liable to over-estimate the carbon benefits of deforestation for growing biofuels, but probably underestimate the other impacts. Therefore the scenarios with afforestation probably underestimate the total radiative forcing and scenarios with deforestation probably overestimate the total radiative forcing.

The RCP and SRES scenarios demonstrate how different cropland policies and changes can affect climate. They represent an unclear future for the land surface's effect on climate, with multiple effects which are subject to many influences. The deliberate action of choosing a pathway that offered bio-geophysical cooling from deforestation for crops could be considered geoengineering and certainly, the effect could be bigger regionally than that of deliberately increasing crop albedo or other crop based geoengineering technology.

6 Increased Crop Albedo

The concept of bio-geoengineering is to produce a cooling effect from increased albedo in crops, without other changes which would accompany land cover changes. It is this, along with the deliberateness of the action, which distinguishes it from other, inadvertent, land cover changes. However, since the change is not between primary plant functional types, the achievable albedo change is likely to be smaller for increased crop albedo than land cover change.

6.1 Albedo Values of Crops

Values of albedo of viable crops are likely to be limited by the natural variability of albedo within a particular crop. Crops have an albedo of around 0.2, similar to grasses. However, records of individual variety leaf and plant albedos are limited, with much of the research having been done many years ago and subject to considerable environmental variability. Therefore more research is needed to establish the range of full spectrum albedos of different crops. It has been suggested from measurements given in the literature that an overall albedo increase in crops of 0.02–0.08 is achievable for crops. The higher end of albedos could prove challenging, but a 0.04 increase is likely to be feasible from conventional breeding using the natural variation in leaf albedo, without the need for genetic modification.[7,6,58] There are bigger differences between different types of crops (*e.g.* between wheat and corn) than within crop types. However, if intra crop substitutions were possible (*i.e.* a lower albedo wheat variety substituted by high albedo wheat) this could increase crop albedo without disrupting food production systems.

6.2   Determinants of Albedo

The overall albedo of a vegetated surface is determined by several aspects. At the leaf level, the albedo is determined by the amounts of reflectance, transmission and absorption at the leaf surface, at different wavelengths (see Figure 2). Reflectance is the fraction of incident radiation reflected by a surface. Transmittance is the fraction of incident light at a specified wavelength that passes through a sample. Because of differences in cell structure affecting light propagation the transmittance is highly variable.[59] Both are expressed as the amount of light as a fraction of the light striking the object.[60]

The albedo of a leaf is affected by not only its colour, caused by chlorophyll levels, but also the leaf wax composition and thickness, trichomes (leaf hairs), the leaf thickness and leaf variegation. Reflective sprays could also be used to increase the albedo at the leaf level.[58] However, at canopy level, leaf albedo is combined with other effects from the canopy morphology; leaf area index; leaf angle distribution; the canopy coverage; the background surface albedo; and the sun zenith angle (see Figure 3).

6.3   Leaf Level Albedo

In general, the spectral characteristics of vegetation are well known due to the use of remote sensing and they have a very high reflection in the near infrared which makes them easily recognisable, see Figure 4(a). Within the leaf,

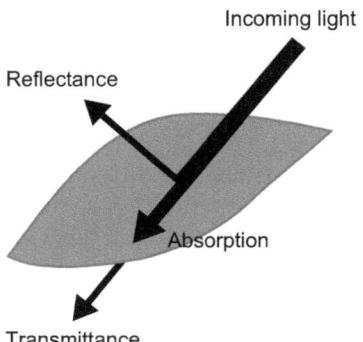

Figure 2   Representation of the potential routes of incoming shortwave radiation at a leaf surface. The light is either reflected away, absorbed by the leaf, or transmitted through the leaf.

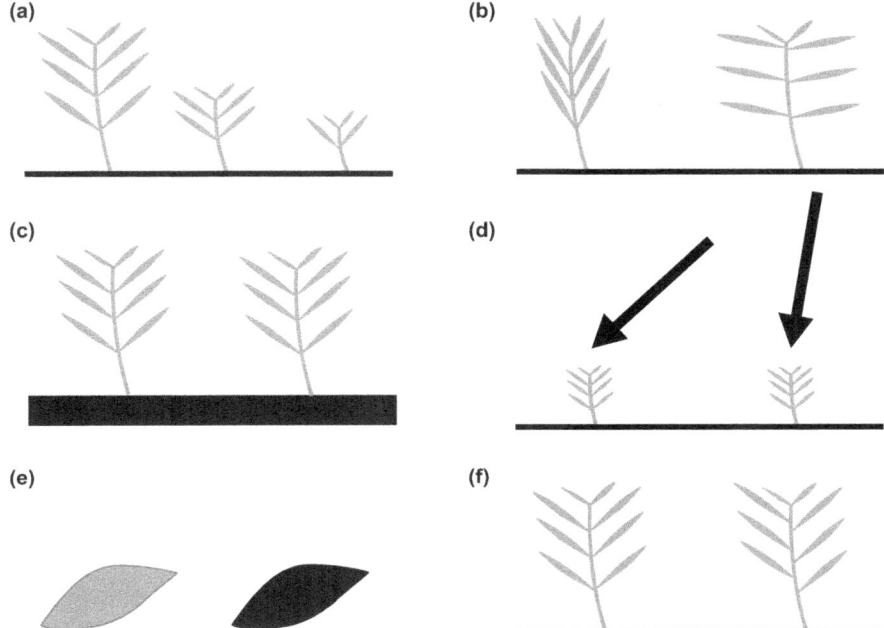

Figure 3 Representation of some the factors influencing crop albedo: (a) leaf area index; (b) leaf angle distribution; (c) background reflectance; (d) solar zenith angle; (e) leaf reflectance; and (f) canopy morphology.

there is low reflectance in the visible blue and red because of high absorption in the photosynthetically active region between 400 and 700 nm.[61] The visible part of the spectrum is also the portion of the incoming solar spectrum with the highest irradiance.[62] A peak in reflectance in the green part of the spectrum gives vegetation its visible colour, see Figure 4(a).

At the leaf level, research has mainly been concentrated on strong identifiable relationships between reflectance at specific wavelengths and plant stress. This allows the use of remote sensing and spectrometers to see areas of sub-optimal crops, and correct issues before they affect yield.[63] The effects of crop health on reflectance are particularly pertinent for bio-geoengineering because a spectral conflict between optimizing yield and optimizing reflectance would be counter-productive.

Although at specific wavelengths there are positive relationships between plant stress and high reflectance, this is not the case across the whole solar spectrum. Nitrogen deficiency increases reflectance at the leaf level because of the negative relationship between low chlorophyll content and reflectance at narrow spectral bands,[63] (at 550 and 700 nm).[64] This relationship is weaker in annual plants, where the leaf structure creates a higher surface reflectance and quickly reaches saturation.[64,65] However, looking across a wider spectrum (from 400 to 750 nm, inclusive) and at larger total chlorophyll

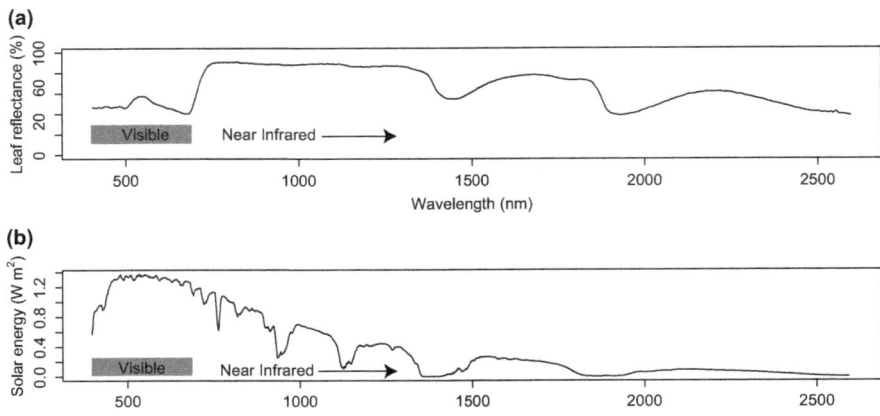

Figure 4   (a) Typical leaf reflectance values (in percent) for wavelengths of 400–2600 nm. (b) Sea level solar spectrum values for wavelengths of 400–2600 nm at sea level, in Watts m^{-2}.

contents, there is a strong statistically significant positive relationship between chlorophyll content and reflection.[66] Since high chlorophyll levels are a direct determinant of potential primary production, increased chlorophyll content to increase albedo could potentially increase yields.

However, the visible part of the spectrum isn't the only factor in determining albedo. The near infrared makes up around 50% of the energy in the solar spectrum at sea level as shown in Figure 4(b), making this a very important range.[67] For instance, red ceramics have a higher albedo than light grey stainless steel, despite being a darker visible colour, because they reflect more strongly in the near infrared.[67] Generally speaking, plant dehydration increases reflectivity at the leaf level in the near infrared.[68–70] At the canopy albedo level this effect is less important due to weaker spectral irradiance in the near infrared.[61,71] Since water absorbs strongly in the near infrared and transmits in the visible, any reduction in water increases the potential for reflection at most wavelengths. This makes the near infrared a useful part of the spectrum for diagnosing plant water content and health but as an albedo increase mechanism, reduced water is likely to have undesirable side effects.

6.4   Canopy Level Albedo

At the canopy level, the leaf level albedo is only a small part of the net resultant albedo. Other factors' influence (see Figure 3) becomes dominant. A key factor in this is how much of the background is covered by the plant. The background soil albedo is especially important when the canopy coverage is low (*e.g.* when the leaf area index is low). Soil albedo is mainly determined by the soil type, water content and soil surface roughness. The largest component of this is the soil colour, which varies with the type of soil, as well as the soil moisture. Dry soil has an albedo up to 0.16 higher than wet soil.[72] This is especially important for crops, as many fields are left uncovered and fallow over winter and ground coverage can be lower than natural land cover because of weed suppression.

The amount of plant coverage of the background soil is mainly determined by the leaf area index, though the net effect on the albedo is not direct. If the background albedo is high, then an increase in leaf area index may reduce the overall albedo. Conversely though, if the background albedo is low, an increase in leaf area index may increase the overall albedo. The leaf area index is dependent on the plant growth stage, but also on the health of the plant. For instance, across the whole canopy in the visible to near infrared, an increase in nitrogen fertilisation (in winter wheat) has the net effect of increasing reflectance at all stages of growth due to increased leaf area index.[65,73] It has even been shown that increased albedo from increased leaf area index and health is currently a small scale summertime impact in areas of North America.[74] Therefore the overall leaf area index and plant health is a significant factor in the overall plant albedo, but is highly dependent on the background albedo.

Similarly, the plant morphology, or architecture, can also affect the albedo, partly through altering the amount of background soil visible. The plant morphology usually refers to the angle and placement of leaves and can vary considerably.[75] Different distributions of leaves and leaf angles lead to different reflectance at the canopy level. Leaf angle distribution (LAD) is a common measure of the orientation of the leaves, which strongly affects the canopy reflectance. The leaf angle distribution is measured from the zenith and thus a high mean leaf angle distribution indicates that the leaves are mainly upright. Different leaf angle distribution values affect what is the most important factor for the total reflectance. An erectophile canopy (mainly vertical leaves) is considerably affected by the background reflectance. For a planophile canopy (mainly horizontal leaves) the background reflectance exerts a much smaller influence and the leaf properties affect the reflectance more.[76] Depending on the background reflectance, canopy reflectance can increase with leaf angle distribution at the red and near infrared parts of the spectrum, *i.e.* more erectophile canopies may have higher canopy reflectance.[77]

Like leaf area index, the leaf angle distribution varies with plant health, development stage and variety. An associated affect of increased nitrogen is a more planophile appearance of the canopy.[78–80] As well as being affected by nitrogen levels, the leaf angle distribution varies naturally between varieties[79,80] and along with the leaf area index is the dominant control on canopy reflectance.[81] Leaf angle distribution values are also seasonally varying. Young plants tend to have more erectophile canopies and varieties tend to be more homogenous in their leaf angle distributions. Differences in both leaf angle distribution and in reflectance appear as the plants develop, resulting in statistically significant differences.[77,76] Therefore differences in reflectance from plant architecture are reliant on not just the variety, but also the growth stage and other factors influencing reflectance. Canopy architecture can also affect the plant health and potentially the crop yield, as planophile leaf angle distributions are also associated with drought resistance.[82]

The albedo is also affected by the angle of the light which falls on the canopy surface. The angle between the horizon and the sun is known as the sun zenith angle. The sun zenith angle changes both diurnally and seasonally, as well as with latitude. It strongly affects the surface albedo because of the changes in scattering angle at the surface from vertical components, especially those with low transmission, such as soil.[83] As the solar zenith angle increases, so does the albedo.[84–86] However, the solar zenith angle effect on albedo is not uniform; cloud cover significantly reduces the effect and am and pm responses differ.[86] Although the solar zenith angle is not a factor which can be deliberately changed to affect crop albedo, it is still an important aspect which needs to be acknowledged. Overall, the key components which can be deliberately manipulated are the background soil and various aspects of the plant health, which can help increase the canopy level albedo, and may even have synergies with maximising yields.

7 Simulations with Climate Models

7.1 Crops in Climate Models

Key questions for the bio-geoengineering concept are how big the cooling effect and what other climatic consequences there might be. These are addressed using climate models. Climate models at their core simulate fundamental physical laws (motion, conservation of mass, *etc.*) across the Earth in a three dimensional grid. Early climate models had a simple representation of the land surface, which used averaged approximations or parameterisations which affect the surface energy balance (shown in Figure 1) such as uniform soil water holding capacity, albedo and roughness length.[87] Later models update the land surface more dynamically (for instance by calculating the restraints on transpiration) and separate out different sections of the land surface (for instance by differentiating between soil and vegetation at the surface, as well as different plant functional types). The newest climate models include new land surface relevant sections such as interactive carbon cycles, dynamic representations of vegetation distribution, higher resolutions and agricultural models. These parts of the model are usually separate to the core physics of the climate model and the model can be run in many different combinations.

Whereas older models either didn't represent different vegetation types or averaged values across a grid box overall, many current models represent vegetation through a tile system. Values for each plant functional type (such as broadleaf or needleleaf trees) are calculated separately and the overall grid box value is calculated according to the proportion of a grid square which is covered by that plant functional type. This has the advantage of providing much better diagnostics about how changes in species composition are affecting the climate. Different plant functional types vary considerably in many parameters, including their temperature ranges, leaf area index ranges and critical humidity deficit. However, by necessity these are wide ranges and many models have only around four to sixteen plant functional types. As a consequence, some land surface models do not explicitly represent agricultural lands or crops. Most crops are physiologically closest to grasses and thus cropland is frequently represented as grass in models without crops specified as a plant functional type.

Thus far, two climate models have been used to perform simulations of increased crop albedo: the Hadley Centre model and the Community Climate System Model. Simulations of an increase in crop albedo of 0.02, 0.04 and 0.08 have been performed with the UK Hadley Centre model, HadCM3, combined with the land surface scheme MOSES2.1[6,7,58] and MOSES1.0.[88] MOSES has only five plant functional types and therefore crops are not explicitly represented. These increased crop albedo simulations use a mask to exclude competition over cropland areas as shown in Figure 5(b). The mask does not allow trees or shrubs to grow in cropland areas, therefore cropland is represented as grass and if the climate is not suitable for the growth of grasses, then the surface is bare soil. The crop albedo was increased by increasing the maximum albedo attainable (calculated with leaf area index) for all the 'crops' within the masked areas.[58] These simulations were run at a range of different carbon dioxide levels, including 280, 350, 560, 700, and 1400 ppm with other initial conditions held at present day levels.[88,58]

Simulations of crop albedo increases of 0.05, 0.10 and 0.15 have also been run with the Community Climate System Model, CCM3.0, using the Community Land Model (CLM3.0).[89] CLM has a more detailed representation of plant functional types and has a separate class for crops, so a crop mask is not needed.[90] These simulations were run at 370 ppm carbon dioxide.[89]

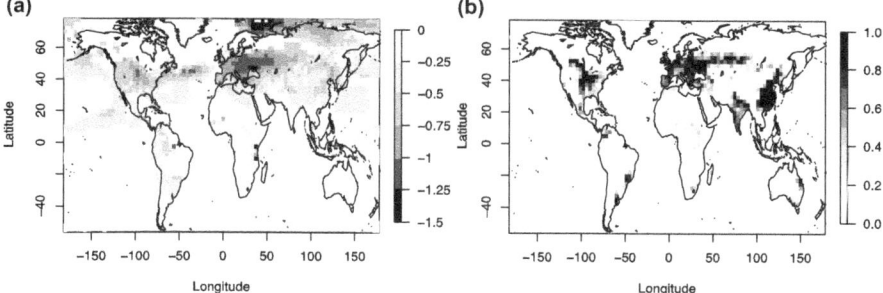

Figure 5   (a) Summertime (June/July/August) statistically significant mean temperature decreases from a 0.04 increase in crop albedo. The simulations shown here have been run to equilibrium at 700 ppm CO$_2$ using HadCM3. Results were tested using a Wilcoxon rank sum test, at 99%. (b) Present day cropland extent, as used for the crop albedo increase in the model.

7.2   Climate Impacts

All the models simulations run so far show that a 0.04 increase in albedo over current cropland areas gives a mean summertime (June, July, August) cooling of around -1 °C, see Figure 5(a). A 0.02 and 0.08 increase in albedo gives a proportional cooling (-0.5 °C and 2 °C, respectively) equating to around 0.25 °C summertime cooling for each 0.01 crop albedo increase. The cooling is concentrated in the temperate regions of albedo increase, with western Europe and North America most affected. The cooling is greatest in the northern hemisphere summertime because of the plants' higher summertime leaf area index which gives greater ground coverage, reducing the influence of the bare soil albedo. In northern hemisphere wintertime the tropical and Asian surface air temperature is strongly affected by the increases in carbon dioxide and thus changes related to the monsoon circulation, which counteract the cooling effect of the increased albedo.[58] Globally, the annual mean cooling is much smaller than the regional seasonal effect: only around -0.11 °C,[7,88,58,6,89] for an increase of 0.04 crop albedo.

Regional patterns of precipitation changes are difficult to predict with climate models, with considerable variability between models,[34] and thus the model results for bio-geoengineering need to be considered with some caution. Precipitation shows a pattern of increase over Europe, with soil moisture showing up a more consistent pattern. The changes in evapo-transpiration give increases in precipitation and also soil moisture over Europe. These changes to temperature and precipitation appear to be robust, as they appear in both HadCM3 and CCSM. Some displaced low latitudes effects give decreased soil moisture in the sub tropics and Australia, which may be causes by changes in cloud formation and precipitation.[89] The results from these models is in contrast to simulations using the climate model NCAR CAM3.1 in which large scale increases in albedo over all land surfaces results in decreased rainfall.[14] How these precipitation changes are affected by the extent and location of increased albedo rather than which model is used is an important issue which has not yet been determined.

In addition to the uncertainties associated with climate model simulations, all of these results are averaged over long periods (30 or 50 years) for climates which are in equilibrium. This enables accurate analysis of the causal links in the simulations, but doesn't reflect the reality of conditions under which bio-geoengineering might be implemented. As bio-geoengineering is a relatively subtle effect it may be difficult to identify at a sub-decadal scale due to high inter-annual variability.

8   Yields

If bio-geoengineering could improve the likelihood of higher crop yields in future, this would be an important outcome. Food security in the form of food production is an essential issue for the 21st Century. Reservations about the potential of crops to provide climate solutions have been voiced on the basis of concerns about reduced yield on the basis that reduced light would reduce photosynthesis.[91] Though this seems intuitively correct, since there is a significant positive correlation of chlorophyll content with increased reflectance, increased reflectance could be beneficial to plant growth. There may also be further advantages in higher albedo crops from the physiological traits used to introduce higher albedo. Trichomes (hairs) and glaucous wax (leaf wax) are both associated with higher yielding varieties in some cases.[92,93]

Further, the cooler summertime climate produced by bio-geoengineering could be advantageous to crop yields and may overcome any negative direct effect, especially in current crop growing regions of Europe and North America, which may become too warm for important crops currently grown. For instance wheat, whose yield is significantly affected by high temperatures and which requires a vernalisation period of cold temperatures in winter (cooling of the seed during germination in order to accelerate flowering when it is planted), is commonly grown in Europe and could be negatively affected by increased temperatures.

There are also potential feedbacks between crop yields, total cropland cover and bio-geoengineering (see Figure 6). The efficacy of bio-geoengineering would be dependent on the extent of crop cover. With more cropland, the cooling effect would be larger. However, if yields were to increase due to bio-geoengineering, fewer crops would be needed for the same demand, potentially implying less bio-geoengineering would be required.

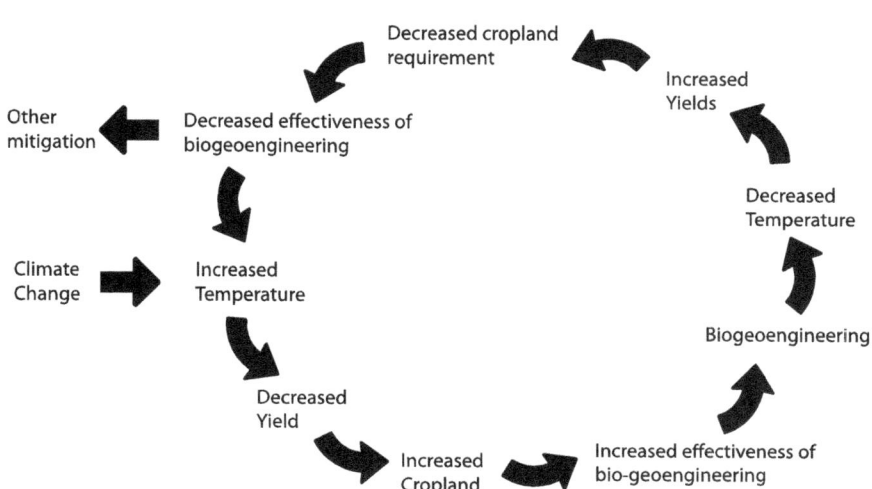

Figure 6 Flow chart of potential interactions between crop yield and biogeoengineering. The cycle begins with climate change and potentially could continue around unless other mitigation solutions could begin to replace biogeoengineering when it began to reduce in efficacy.

This could result in a negative feedback loop. The most feasible way out of this feedback loop would be through other climate change mitigation solutions (see Figure 6). This emphasizes that increased crop albedo is inherently a geoengineering solution which must be used in conjunction with conventional mitigation.

9 Other Crop Cooling Potential

Crops not only have potential for bio-geoengineering but also as a carbon capture and storage geoengineering method and a more conventional carbon reduction and mitigation technique. Many of these methods could be synergistic with increased crop albedo and together achieve significant cooling.

9.1 Soil Carbon Sequestration

Soil carbon currently makes up around 80% of the terrestrial carbon store.[94] However, soil degradation through land use change can release 25–30% of this soil carbon into the atmosphere.[95] If soil carbon sequestration in degraded cropland could be increased, enhancing the removal of carbon dioxide from the atmosphere, it could additionally take up 0.4–1.2 gigatonnes of carbon each year,[96] although this is regionally variable.[97] This is equivalent to up to 15% of fossil fuel emissions and could be a substantial contribution to reducing carbon emissions.[96] Enhancing carbon removal into soil carbon stores could also have some synergies with increasing crop yield, because of the increased soil organic carbon which is a store of nutrients. Agricultural management policies which deliberately increase soil carbon sequestration could be considered a type of carbon capture and storage geoengineering.

9.2 Biofuels

Most crops are utilized for direct or indirect food purposes, but crops can also be used for textiles, fuel and other uses. Crops which are processed into fuel, especially fuel for vehicles, are known as biofuel crops. Being a renewable source of energy, biofuels have been considered an important way of reducing fossil fuel use and are included explicitly in the RCP scenarios. By replacing fossil-fuels, biofuels avoid the emissions which would be emitted and provide a sustainable power source. However, biofuels have gathered criticism for driving up food prices and causing indirect land use change.[98–101] Moreover, care must be taken to consider the whole production process since increased use of fertilizers (which currently require considerable carbon emissions in their production as well as emitting the greenhouse gas nitrous oxide) on biofuels could negate the positive carbon impact.[102]

10 Priorities for Future Work

There are currently important gaps in our knowledge about how bio-geoengineering could work. Questions remain in the plant physiological, climatic and practical implementation areas.

One of the most important feasibility issues is what range of higher albedo crops could be bred. Finding out what the existing variability within crops is and seeing if it is sufficient to increase crop albedo using conventional breeding methods is a crucial part of this. If there is not a large, controllable natural variability which could be selectively bred for, would genetic modification be a viable and acceptable way to create high albedo crops? However, if the physiological traits which provide the greatest increase in albedo could be identified and conventionally bred for, then no genetic engineering would be necessary. Further, there may be win-wins between increased albedo, plant health and increased drought resistance or other desirable traits, which would be worth exploring.

From a climatic perspective, it is important that simulations of increased crop albedo are run with more climate models, as two models are unlikely to capture the whole of the potential consequences. Multiple models could also be used to understand how varying levels of implementation (across spatial and temporal scales) would affect the mean climate and the climate variability. Using transient simulations which include bio-geoengineering could aid seeing how bio-geoengineering could work in synergy with other mitigation policies. The extent to which seasonal variability is affected by cropping cycles would also need to be addressed, as the climate models used so far have no parameterization of this, which may affect the wintertime climate, when few crops are being grown. Similarly, the question of what could be the impact on yield from the changes in climate and from the physiological changes to crop varieties is a complicated issue which must be addressed before bio-geoengineering can be considered to be feasible. The answers to these questions are not necessarily straightforward and in some cases there are no strong precedents on how to do this research.

11 Conclusions

Cooling with crops could provide local help towards managing and mitigating against the most harmful aspects of climate change, but is a regional help, rather than a global solution. Future climate and agriculture are inextricably linked; yields determine how much land will need to be converted to cropland, whilst the climate change which is partially determined by the land surface effects will affect crop yields.

Food security is one of the most crucial challenges for the 21st Century and making crops part of the solution can be perceived as a risk. However, the opinion that some things are 'too important' to make concessions to climate change is based on a false assessment of risk and the interconnectedness of these systems. Excluding crops and other important factors as potential options for reducing the impact of climate change will leave the necessity of a single, large solution. If that one solution fails, then there is the potential for a larger problem. The least risk strategy is actually to draw together many small changes to help combat climate change, spreading the risk. Similarly, criticisms of bio-geoengineering on the basis that it doesn't contribute enough cooling miss the essential point: as a multifaceted problem, climate change needs a multifaceted solution. Doing nothing about climate change is likely to result in substantial and damaging changes to the earth's climate; cooling with crops could be one of many building blocks which creates a whole solution.

References

1. *World Bank Indic. Databank.*
2. P. Smith, P. J. Gregory, D. van Vuuren, M. Obersteiner, P. Havlík, M. Rounsevell, J. Woods, E. Stehfest and J. Bellarby, *Philos. Trans. R. Soc., Ser. B.*, 2010, **365**, 2941–2957.
3. S. K. Rose, H. Ahammad, B. Eickhout, B. Fisher, A. Kurosawa, S. Rao, K. Riahi and D. P. van Vuuren, *Energy Econ.*, 2012, **34**, 365–380.
4. M. Fowles, *Biomass Bioenergy*, 2007, **31**, 426–432.
5. J. Lehmann, J. Gaunt and M. Rondon, *Mitigat. Adapt. Strat. Glob. Change J.*, 2006, **11**, 395–419.
6. J. S. Singarayer and T. Davies-Barnard, *Philos. Trans. R. Soc. Math. Phys. Eng. Sci.*, 2012, **370**, 4301–4316.
7. A. Ridgwell, J. S. Singarayer, A. M. Hetherington and P. J. Valdes, *Curr. Biol.*, 2009, **19**, 146–150.
8. J. O. Kaplan, K. M. Krumhardt and N. Zimmermann, *Quat. Sci. Rev.*, 2009, **28**, 3016–3034.
9. Y. Pan, R. A. Birdsey, J. Fang, R. Houghton, P. E. Kauppi, W. A. Kurz, O. L. Phillips, A. Shvidenko, S. L. Lewis, J. G. Canadell, P. Ciais, R. B. Jackson, S. W. Pacala, A. D. McGuire, S. Piao, A. Rautiainen, S. Sitch and D. Hayes, *Science*, 2011, **333**, 988–993.
10. O. A. Anisimov, *Environ. Res. Lett.*, 2007, **2**, 045016.
11. G. B. Bonan, *Science*, 2008, **320**, 1444–1449.
12. L. Breuer, K. Eckhardt and H.-G. Frede, *Ecol. Model.*, 2003, **169**, 237–293.
13. H. H. G. Savenije, *Hydrol. Process.*, 2004, **18**, 1507–1511.
14. G. Bala and B. Nag, *Clim. Dyn.*, 2012, **39**, 1527–1542.
15. M. Jin and S. Liang, *J. Clim.*, 2006, **19**, 2867–2881.
16. R. H. Shaw and A. Pereira, *Agric. Meteorol.*, 1982, **26**, 51–65.
17. R. E. Dickinson and P. Kennedy, *Geophys. Res. Lett.*, 1992, **19**, 1947–1950.
18. N. Gedney and P. J. Valdes, *Geophys. Res. Lett.*, 2000, **27**, 3053–3056.
19. G. F. Pires and M. H. Costa, *Geophys. Res. Lett.*, 2013, **40**, 3618–3623.
20. P. K. Snyder, *Earth Interactions*, 2010, **14**, 1–34.

21. J. Shukla and Y. Mintz, *Science*, 1982, **215**, 1498–1501.
22. E. a. B. Eltahir and R. L. Bras, *Q. J. R. Meteorol. Soc.*, 1994, **120**, 861–880.
23. D. C. Nepstad, C. R. de Carvalho, E. A. Davidson, P. H. Jipp, P. A. Lefebvre, G. H. Negreiros, E. D. da Silva, T. A. Stone, S. E. Trumbore and S. Vieira, *Nature*, 1994, **372**, 666–669.
24. L. E. O. C. Aragão and Y. E. Shimabukuro, *Science*, 2010, **328**, 1275–1278.
25. S. J. Wright, *Ann. N. Y. Acad. Sci.*, 2010, **1195**, 1–27.
26. L. J. C. Oliveira, M. H. Costa, B. S. Soares-Filho and M. T. Coe, *Environ. Res. Lett.*, 2013, **8**, 024021.
27. J. Lean and P. R. Rowntree, *J. Clim.*, 1997, **10**, 1216–1235.
28. R. A. Betts, *Nature*, 2000, **408**, 187–190.
29. R. E. Leonard and A. R. Eschner, *Water Resour. Res.*, 1968, **4**, 931–935.
30. J. P. Boisier, N. de Noblet-Ducoudré and P. Ciais, *Biogeosciences*, 2013, **10**, 1501–1516.
31. W. F. Ruddiman, *Annu. Rev. Earth Planet. Sci.*, 2013, **41**, 45–68.
32. V. Brovkin, M. Claussen, E. Driesschaert, T. Fichefet, D. Kicklighter, M. F. Loutre, H. D. Matthews, N. Ramankutty, M. Schaeffer and A. Sokolov, *Clim. Dyn.*, 2006, **26**, 587–600.
33. N. de Noblet-Ducoudré, J.-P. Boisier, A. Pitman, G. B. Bonan, V. Brovkin, F. Cruz, C. Delire, V. Gayler, B. J. J. M. van den Hurk, P. J. Lawrence, M. K. van der Molen, C. Müller, C. H. Reick, B. J. Strengers and A. Voldoire, *J. Clim.*, 2012, **25**, 3261–3281.
34. S. Solomon, D. Qin, M. Manning, Z. Chen, M. Marquis, K. B. Avery, M. Tignor and H. L. Miller, Eds., *IPCC, 2007: Climate Change 2007: The Physical Science Basis. Contribution of Working Group I to the Fourth Assessment Report of the Intergovernmental Panel on Climate Change*, Cambridge University Press, Cambridge, UK, 2007.
35. J. Pongratz, C. H. Reick, T. Raddatz and M. Claussen, *Geophys. Res. Lett.*, 2010, **37**, L08702.
36. H. D. Matthews, A. J. Weaver, K. J. Meissner, N. P. Gillett and M. Eby, *Clim. Dyn.*, 2004, **22**, 461–479.
37. V. Brovkin, S. Sitch, W. Von Bloh, M. Claussen, E. Bauer and W. Cramer, *Glob. Change Biol.*, 2004, **10**, 1253–1266.
38. D. Nepstad, B. S. Soares-Filho, F. Merry, A. Lima, P. Moutinho, J. Carter, M. Bowman, A. Cattaneo, H. Rodrigues, S. Schwartzman, D. G. McGrath, C. M. Stickler, R. Lubowski, P. Piris-Cabezas, S. Rivero, A. Alencar, O. Almeida and O. Stella, *Science*, 2009, **326**, 1350–1351.
39. J. P. Malingreau, H. D. Eva and E. E. de Miranda, *AMBIO*, 2012, **41**, 309–314.
40. E. Y. Arima, P. Richards, R. Walker and M. M. Caldas, *Environ. Res. Lett.*, 2011, **6**, 024010.
41. J. C. Jiménez-Muñoz, J. A. Sobrino, C. Mattar and Y. Malhi, *J. Geophys. Res. Atmos.*, 2013, **118**, 5204–5215.
42. P. Meiyappan and A. K. Jain, *Front. Earth Sci.*, 2012, **6**, 122–139.

43. J. Feddema, K. Oleson, G. Bonan, L. Mearns, W. Washington, G. Meehl and D. Nychka, *Clim. Dyn.*, 2005, **25**, 581–609.
44. N. W. Arnell, M. J. L. Livermore, S. Kovats, P. E. Levy, R. Nicholls, M. L. Parry and S. R. Gaffin, *Glob. Environ. Change*, 2004, **14**, 3–20.
45. J. J. Feddema, K. W. Oleson, G. B. Bonan, L. O. Mearns, L. E. Buja, G. A. Meehl and W. M. Washington, *Science*, 2005, **310**, 1674–1678.
46. M. Meinshausen, S. Smith, K. Calvin, J. Daniel, M. Kainuma, J.-F. Lamarque, K. Matsumoto, S. Montzka, S. Raper, K. Riahi, A. Thomson, G. Velders and D. P. van Vuuren, *Clim. Change*, 2011, **109**, 213–241.
47. R. H. Moss, J. A. Edmonds, K. A. Hibbard, M. R. Manning, S. K. Rose, D. P. van Vuuren, T. R. Carter, S. Emori, M. Kainuma, T. Kram, G. A. Meehl, J. F. B. Mitchell, N. Nakicenovic, K. Riahi, S. J. Smith, R. J. Stouffer, A. M. Thomson, J. P. Weyant and T. J. Wilbanks, *Nature*, 2010, **463**, 747–756.
48. G. Hurtt, L. Chini, S. Frolking, R. Betts, J. Feddema, G. Fischer, J. Fisk, K. Hibbard, R. Houghton, A. Janetos, C. Jones, G. Kindermann, T. Kinoshita, K. Klein Goldewijk, K. Riahi, E. Shevliakova, S. Smith, E. Stehfest, A. Thomson, P. Thornton, D. van Vuuren and Y. Wang, *Clim. Change*, 2011, **109**, 117–161.
49. V. Brovkin, L. Boysen, V. K. Arora, J.-P. Boisier, P. Cadule, L. Chini, M. Claussen, P. Friedlingstein, V. Gayler, J. J. M. van der Hurk, G. C. Hurtt, C. D. Jones, E. Kato, N. de Noblet-Ducoudré, F. Pacifico, J. Pongratz and M. Weiss, *J. Clim.*, 2013, **26**, 6859–6881.
50. J. Bruinsma, *World Agriculture: Towards 2015/2030: An FAO Perspective*, Earthscan/James & James, London, 2003.
51. J. Gornall, R. Betts, E. Burke, R. Clark, J. Camp, K. Willett and A. Wiltshire, *Philos. Trans. R. Soc., Ser. B.*, 2010, **365**, 2973–2989.
52. K. Riahi, S. Rao, V. Krey, C. Cho, V. Chirkov, G. Fischer, G. Kindermann, N. Nakicenovic and P. Rafaj, *Clim. Change*, 2011, **109**, 33–57.
53. A. M. Thomson, K. V. Calvin, S. J. Smith, G. P. Kyle, A. Volke, P. Patel, S. Delgado-Arias, B. Bond-Lamberty, M. A. Wise, L. E. Clarke and J. A. Edmonds, *Clim. Change*, 2011, **109**, 77–94.
54. D. van Vuuren, J. Edmonds, M. Kainuma, K. Riahi, A. Thomson, K. Hibbard, G. Hurtt, T. Kram, V. Krey, J.-F. Lamarque, T. Masui, M. Meinshausen, N. Nakicenovic, S. Smith and S. Rose, *Clim. Change*, 2011, **109**, 5–31.
55. T. Davies-Barnard, C. Jones, P. J. Valdes and J. S. Singarayer, *J. Clim*, 2014, **27**, 1413–1424.
56. D. M. Lapola, R. Schaldach, J. Alcamo, A. Bondeau, J. Koch, C. Koelking and J. A. Priess, *Proc. Natl. Acad. Sci. U. S. A.*, 2010, **107**, 3388–3393.
57. M. Wise, K. Calvin, A. Thomson, L. Clarke, B. Bond-Lamberty, R. Sands, S. J. Smith, A. Janetos and J. Edmonds, *Science*, 2009, **324**, 1183–1186.
58. J. S. Singarayer, A. Ridgwell and P. Irvine, *Environ. Res. Lett.*, 2009, **4**, 045110.
59. J. Otterman, T. Brakke and J. Smith, *Remote Sens. Environ.*, 1995, **54**, 49–60.

60. J. T. Woolley, *Plant Physiol.*, 1971, **47**, 656–662.
61. H. W. Gausman, *Remote Sens. Environ.*, 1977, **6**, 1–9.
62. T. W. Mulroy, *Oecologia*, 1979, **38**, 349–357.
63. G. A. Carter and A. K. Knapp, *Am. J. Bot.*, 2001, **88**, 677–684.
64. D. A. Sims and J. A. Gamon, *Remote Sens. Environ.*, 2002, **81**, 337–354.
65. I. Filella, L. Serrano, J. Serra and J. Peñuelas, *Crop Sci.*, 1995, **35**, 1400–1405.
66. A. A. Gitelson, Y. Gritz and M. N. Merzlyak, *J. Plant Physiol.*, 2003, **160**, 271–282.
67. R. T. A. Prado and F. L. Ferreira, *Energy Build*, 2005, **37**, 295–300.
68. G. A. Carter, *Am. J. Bot.*, 1993, **80**, 239–243.
69. E. Hunt and B. Rock, *Remote Sens. Environ.*, 1989, **30**, 43–54.
70. C. J. Tucker, *Remote Sens. Environ.*, 1980, **10**, 23–32.
71. C. A. Gueymard, *Sol. Energy*, 2004, **76**, 423–453.
72. S. B. Idso, R. D. Jackson, R. J. Reginato, B. A. Kimball and F. S. Nakayama, *J. Appl. Meteorol.*, 1975, **14**, 109–113.
73. L. D. Hinzman, M. E. Bauer and C. S. T. Daughtry, *Remote Sens. Environ.*, 1986, **19**, 47–61.
74. D. B. Lobell, G. Bala and P. B. Duffy, *Geophys. Res. Lett.*, 2006, **33** L06708.
75. D. S. Falster and M. Westoby, *New Phytol.*, 2003, **158**, 509–525.
76. S. R. Phinn, D. A. Stow and D. Van Mouwerik, *Photogramm. Eng. Remote Sens.*, 1999, **65**, 485–493.
77. W. Huang, Z. Niu, J. Wang, L. Liu, C. Zhao and Q. Liu, *IEEE Trans. Geosci. Remote Sens.*, 2006, **44**, 3601–3609.
78. N. Broge and J. Mortensen, *Remote Sens. Environ.*, 2002, **81**, 45–57.
79. R. D. Jackson and P. J. Pinter Jr., *Remote Sens. Environ.*, 1986, **20**, 43–56.
80. P. M. Hansen and J. K. Schjoerring, *Remote Sens. Environ.*, 2003, **86**, 542–553.
81. G. P. Asner, *Remote Sens. Environ.*, 1998, **64**, 234–253.
82. C. Werner, R. J. Ryel, O. Correia and W. Beyschlag, *Acta Oecologica*, 2001, **22**, 129–138.
83. D. S. Kimes, *Appl. Opt.*, 1983, **22**, 1364–1372.
84. F. Yang, K. Mitchell, Y.-T. Hou, Y. Dai, X. Zeng, Z. Wang and X.-Z. Liang, *J. Appl. Meteorol. Clim.*, 2008, **47**, 2963–2982.
85. Z. Wang, M. Barlage, X. Zeng, R. E. Dickinson and C. B. Schaaf, *Geophys. Res. Lett.*, 2005, **32**, n/a–n/a.
86. L. C. Nkemdirim, *J. Appl. Meteorol.*, 1972, **11**, 867–874.
87. A. J. Pitman, *Int. J. Clim.*, 2003, **23**, 479–510.
88. P. J. Irvine, A. Ridgwell and D. J. Lunt, *J. Geophys. Res. Atmos.*, 2011, **116**, L18702.
89. C. E. Doughty, C. B. Field and A. M. S. McMillan, *Clim. Change*, 2011, **104**, 379–387.
90. P. J. Lawrence and T. N. Chase, *J. Geophys. Res. Biogeosci.*, 2007, **112**, G01023.

91. N. E. Vaughan and T. M. Lenton, *Clim. Change*, 2011, **109**, 745–790.
92. A. Febrero, S. Fernández, J. L. Molina-Cano and J. L. Araus, *J. Exp. Bot.*, 1998, **49**, 1575–1581.
93. D. A. Johnson, R. A. Richards and N. C. Turner, *Crop Sci.*, 1983, 23, 318–325.
94. H. H. Janzen, *Agric. Ecosyst. Environ.*, 2004, **104**, 399–417.
95. R. A. Houghton, *Tellus B*, 2010, **62**, 337–351.
96. R. Lal, *Science*, 2004, **304**, 1623–1627.
97. S. K. Lam, D. Chen, A. R. Mosier and R. Roush, *Sci. Rep.*, 2013, 3.
98. A. Ajanovic, *Energy*, 2011, **36**, 2070–2076.
99. D. Headey and S. Fan, *Agric. Econ.*, 2008, **39**, 375–391.
100. T. Searchinger, R. Heimlich, R. A. Houghton, F. Dong, A. Elobeid, J. Fabiosa, S. Tokgoz, D. Hayes and T.-H. Yu, *Science*, 2008, **319**, 1238–1240.
101. R. J. Plevin, M. O'Hare, A. D. Jones, M. S. Torn and H. K. Gibbs, *Environ. Sci. Technol.*, 2010, **44**, 8015–8021.
102. P. J. Crutzen, A. R. Mosier, K. A. Smith and W. Winiwarter, *Atmos. Chem. Phys.*, 2008, **8**, 389–395.

Engineering Ideas for Brighter Clouds

STEPHEN H. SALTER,* THOMAS STEVENSON AND ANDREAS TSIAMIS

ABSTRACT

It may be possible to reduce global warming by increasing the reflectivity of marine stratocumulus clouds thereby reducing the amount of solar energy that is absorbed. Quite a small change to the reflectivity could stop further temperature rise or even produce a reversion towards pre-industrial values. This paper gives a brief account of the physics behind the Twomey effect and its application for marine cloud brightening by the release of sub-micron drops of sea water into the marine boundary layer using a fleet of mobile spray vessels. We argue that the mobility of spray vessels and the short life of spray are advantageous by allowing rapid tactical control in response to local conditions. We identify the main engineering problem as spray production, which in turn requires ultra-filtration of plankton-rich seawater. The proposed engineering solutions involving Rayleigh nozzles etched in silicon and piezo-electric excitation are illustrated with drawings. The results of a COMSOL Multiphysics simulation of drop generation are given, with nozzle diameter, drive pressure, excitation frequency and power requirement as functions of drop diameter. The predicted power requirement is higher than initially hoped for and this has led to a modified vessel design with active hydrofoils giving much lower drag than displacement hulls and turbines. The active control of hydrofoil pitch angle can be used for power generation, roll stabilizing and may also reduce hull loading similarly to the suspension systems of road vehicles. The need to identify unwanted side effects of marine

*Corresponding author

Issues in Environmental Science and Technology, 38
Geoengineering of the Climate System
Edited by R.E. Hester and R.M. Harrison
© The Royal Society of Chemistry 2014
Published by the Royal Society of Chemistry, www.rsc.org

cloud brightening has led to a method for using climate models to give an everywhere-to-everywhere transfer function of the effects of spray in each region on weather records at all observing stations. The technique uses individual coded modulation of the concentration of cloud-condensation nuclei separately in each of many spray regions and is based on methods used for small-signal detection in electronic systems. The first use in a climate model shows very accurate measurement of changes to a temperature record and that that marine cloud brightening can affect precipitation in *both* directions. Replication with other climate models will be necessary. The paper ends with tentative estimates for the cost of mass production spray vessels based on actual quotations for parts of the spray generation hardware and on the cost of Flower-class corvettes used by the Royal Navy in World War II which were built in similar numbers.

1   Introduction

Any reader of this paper will already know about increases in atmospheric CO_2, Arctic ice loss, methane release, carbon embedded in imports and the progress to date of our world leaders in finding solutions to these problems. This paper describes some of the engineering ideas needed to implement a proposal by John Latham to increase the reflectivity of marine stratocumulus clouds by an amount necessary to offset the thermal effects of increased greenhouse gases.[1]

2   A Reminder of the Physics

The power density of the solar input at the top of the atmosphere, not quite constant, is about 1360 W m^{-2}. At mid-latitudes the input over 24 hours is about 340 W m^{-2}. Changes since preindustrial times have retained about 1.6 W m^{-2} more than before. If CO_2 concentrations are to double, the extra power density is expected to be about 3.7 W m^{-2}, which is less than 1.1% of the input. Quite a small change to the reflectivity of the earth or its clouds could stop further temperature rises or even produce a reversion to preindustrial values.

The most commonly mentioned method to increase the reflectivity of the earth is the injection of aerosol particles such as SO_2 into the stratosphere as discussed by Robock in Chapter 7. In this chapter we discuss the engineering design for a proposal for the use of sub-micron drops of filtered sea water. Drops released from near the surface would be spread by turbulence through the marine boundary layer. We can make an engineering estimate of turbulence by taking about 15 minutes of video of marine cloud formations and speeding it up with a viewer which allows continuous scrolling back and forth through the sequence.[2]

The speeding up lets us see that clouds behave like floppy rollers with diameters reaching from sea level to cloud top of which only the top part of the roller becomes visible as increasing height produces cooling to the point of condensation. We can see that 180° of roller rotation takes about 10 min indicating velocities up and down of the order of 1 m s^{-1}. Nature does not like uneven concentrations and uses turbulence to spread nuclei fairly evenly through the boundary layer.

After release near sea level the drops will evaporate to leave crystals of dry salt. The ratio of drop diameter from 3.5% salinity sea water to a dry salt sphere is 3.92. The ratio to the side of a dry salt cube is 4.86. Dry salt has a high reflectivity and some solar energy will be reflected back to space. This initial gain in reflection from the dry crystals is called the 'direct Twomey effect'. However, there is also a second mechanism known as the 'indirect Twomey effect' which occurs if the salt crystal reaches the cloud.[3] Even if the relative humidity in air is above 100% a drop cannot form without a nucleus to start its growth. In air over typical land there are 1000 to 5000 nuclei cm^{-3} and so drops form very close to the spout of a boiling kettle. In the clean air of the mid ocean, however, there may be only 10 to 100 nuclei cm^{-3} and so the water that cannot be in vapour form has to be in relatively large drops, with diameters of the order of 25 μm.[4]

Hydrophilic materials like sea salt of the right size are excellent cloud condensation nuclei. If an extra nucleus approaches a 25 μm drop, water can evaporate from the larger nucleus and condense on the smaller to produce the same liquid volume in two drops each 19.84 μm in diameter. The ratio of projected areas rises by 26%. In some conditions, particles in ship exhaust gases can increase the reflectivity of marine stratocumulus clouds enough to be detected by eye, around 20%. Twomey used cloud reflectivity observations from satellites and nuclei concentration from aircraft to investigate the effect. Schwarz and Slingo derived an analytical equation for reflectivity change based on cloud depth, liquid water content and the initial concentration of cloud condensation nuclei.[5] For thin clouds and common ranges of other parameters, the change in reflectivity is 0.075 of the natural log of the fractional change in the concentration of condensation nuclei. If $N1$ is the initial drop concentration and $N2$ the drop concentration after spray the change in cloud reflectivity is

$$\Delta R = 0.075 \ln\left(\frac{N2}{N1}\right)$$

This means that a doubling of the number of nuclei will increase cloud reflectivity by 0.058 from a typical value of 0.5. Latham showed that the volumes of water which would have to be sprayed to reverse the thermal effects of anthropogenic damage since pre-industrial times were surprisingly small, of the order of 10 m^3 s^{-1} for 0.8 μm drops.[1]

The calculation depends on assumptions on nucleus life and initial nuclei concentration. The life of the nuclei is shortened by rain and drizzle. Smith

Park and Conserdine give a graph as a function of drop size suggesting a typical half-life of 60 h for our size.[6] The assumed initial nuclei concentrations are the ones suggested by Bennartz for clean mid ocean air.[7] The short life means that the spraying process must be continuous but offers the option of rapid, tactical control which can be varied to suit satellite observations of raised sea surface temperatures, the phase of monsoons and the state of the El Niño oscillation. Control engineers will appreciate the low phase lag. A short life also allows us to avoid getting any aerosol over the Arctic in winter where it would act as a blanket to reflect back long wave radiation going out to space as studied by Kristjanssen.[8]

Cloud reflectivity and the resulting energy changes can also be predicted by global climate models. However, the best modellers are quick to point out that agreement between different climate models is not good. Some of the differences can be explained by the differences in assumed values of drop life, initial nuclei concentration and the spread of spray diameters which not always specified by modellers. The changes in cloud reflectivity needed to reverse global warming are well below what can be detected by eye. However, it may be possible to superimpose large numbers of satellite images to enhance the contrast between a single spray wake and the surrounding clouds so as to measure effects in a wide range of climate conditions and geographical regions.

3   The Main Engineering Problems

We need an energy-efficient mechanism to produce a mono-disperse, submicron spray despite the plankton, oil and silt found in sea water. We need mobile platforms which can generate energy, be moved round the world to suit tactical spray plans and have acceptable, if not total chance, of surviving extreme conditions. We need fairly long and well-matched service intervals, at least as long as the intervals for antifouling treatment of ship hulls. We should avoid the need for any appreciable volume of consumable materials which cannot be made at sea. However, we could make at least chlorine, hydrochloric and nitric acid, ammonia, sodium hydroxide, sodium carbonate, ozone and hydrogen peroxide.

3.1   Spray Generation

The most challenging problem in the entire project is the production of spray. After consideration of spinning disks, electrostatic bagatelle, the high velocity collision of opposed jets saturated with high pressure air and Taylor cones produced by high voltage fields, we settled on the well-known technique studied by Rayleigh of pumping water through small nozzles but with high-frequency ultrasonic excitation. Neukermans describes work on the expansion of supercritical salt water through much larger nozzles.[9]

The nearest present technology for the spray generation is inkjet printing. Several eminent pioneers in the ink jet industry were consulted. Their opinions were unanimous and emphatic that the nozzle clogging problem was totally insoluble. It is therefore useful to identify differences between the two requirements. The drop diameter of the very best graphic arts ink jet printers of 2013 is 15 μm and the suggested diameter for this project is 0.8 μm, so the ratio of drop masses is 6600.

However, while a single blocked inkjet nozzle will spoil the look of text, a billion blocked nozzles will reduce the output of a spray vessel by only 2%. Despite having sticky pigments, an inkjet nozzle must operate first time after months on the shelf and weeks of inaction but the ink on the paper must be dry enough to be handled in few seconds. Spray nozzles work with no solid content, and can have an elaborate start-up and shut-down procedure. They can be back-flushed with fresh water every few minutes and dried with ultra clean air. Inkjet parts must sell for a few pounds, weigh a few tens of grams and manage with no filtration after leaving the factory. The filters for a spray vessel can weigh more than a tonne, operate continuously and form an essential and critical part of a £2 million vessel.

The COMSOL simulation shows that mono-disperse drops of the right size can be produced. Drop regularity is aided by a small amount of ultrasonic excitation. For 800 nm drops we need a 370 nm nozzle, a pressure of 80 bar and an excitation frequency of 27 MHz as shown in Figure 1.

The predictions for the pressure needed to make drops are in reasonable agreement with an equation given by van Hoeve *et al.*, who write that the lowest critical velocity can be expressed in terms of a Weber number.[10] The Weber number is the ratio of kinetic to surface energy of drops in a jet. If ρ is fluid density, d is jet diameter, U is jet velocity and γ is the surface tension the Weber number is

$$\text{We} = \frac{\rho d U^2}{\gamma}$$

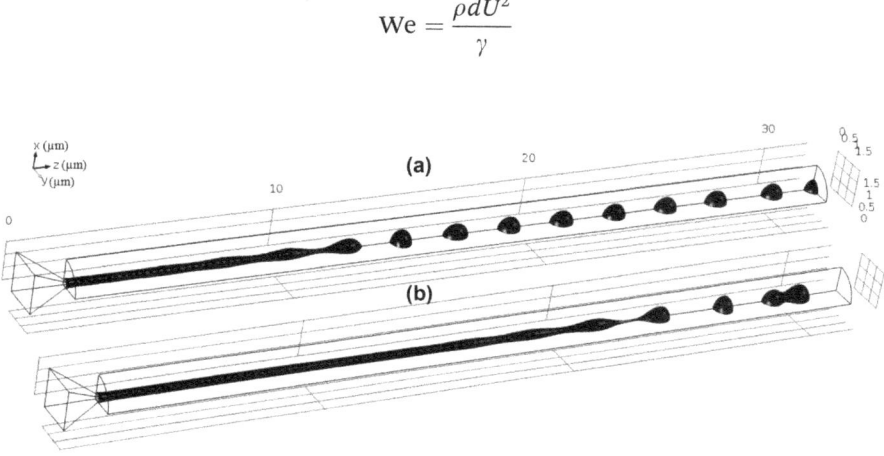

Figure 1 A COMSOL Multiphysics simulation showing that a small amount of pressure modulation at the frequency predicted by Rayleigh will enhance drop breakup and narrow the spread of drop diameters.

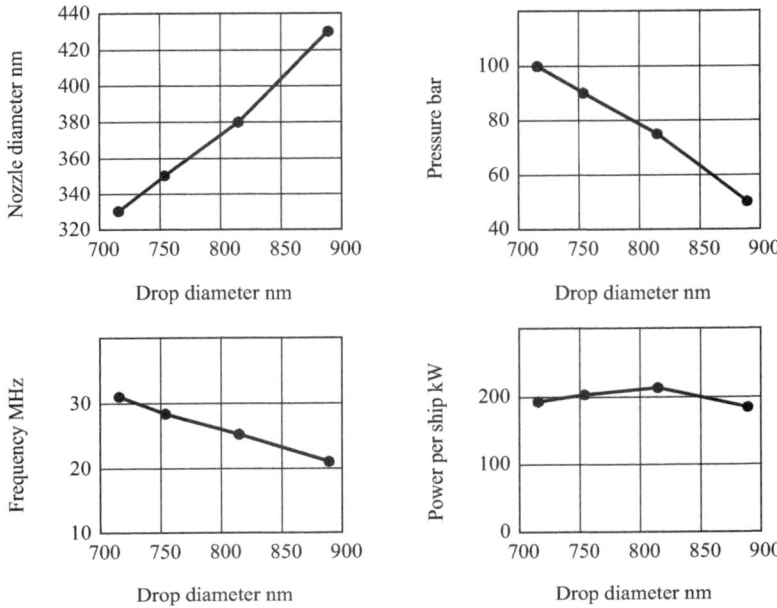

Figure 2   COMSOL results for 'nice' drop breakup showing the values for nozzle diameter, pressure and excitation frequency which are suitable for a range of drop diameters. The power is based on the pumping to produce 10^{17} drops s^{-1} per ship, not including the drive for piezo-electrics. It is the drop number of the best size that drives the Twomey effect, not spray volume. This leads to the flatness of the power *vs.* diameter curve.

This must be >8 for good drop breakup. We can replace the square of velocity with 2 times pressure over liquid density, giving an absolute minimum pressure of 4 times surface tension over nozzle diameter and then add more pressure for the pressure drop in the approach to the nozzle. COMSOL predicts that, with a small pressure pulsation at the best frequency, drops show some ellipsoidal wobbles and then settle to a spherical shape. The COMSOL Multiphysics software has been used to show in Figure 2 how the choice of drop diameter affects the values for pressure, nozzle size and excitation frequency.

The popular choice of piezo materials for high power and high frequency is PZT4. Piezo electric materials have a very high dielectric constant and so an element with a fundamental resonance at 27 MHz would have a very large capacitance needing enormous currents. We can reduce the problem by using thicker ceramics and driving them at a high harmonic, we hope the tenth, of the fundamental. At these frequencies the magnetic fields around a conductor force the current away from the centre and all the low frequency rules for resistance are wrong. We mount two nozzle wafers back-to-back as close as possible, as shown in Figure 3, and send current back and forth

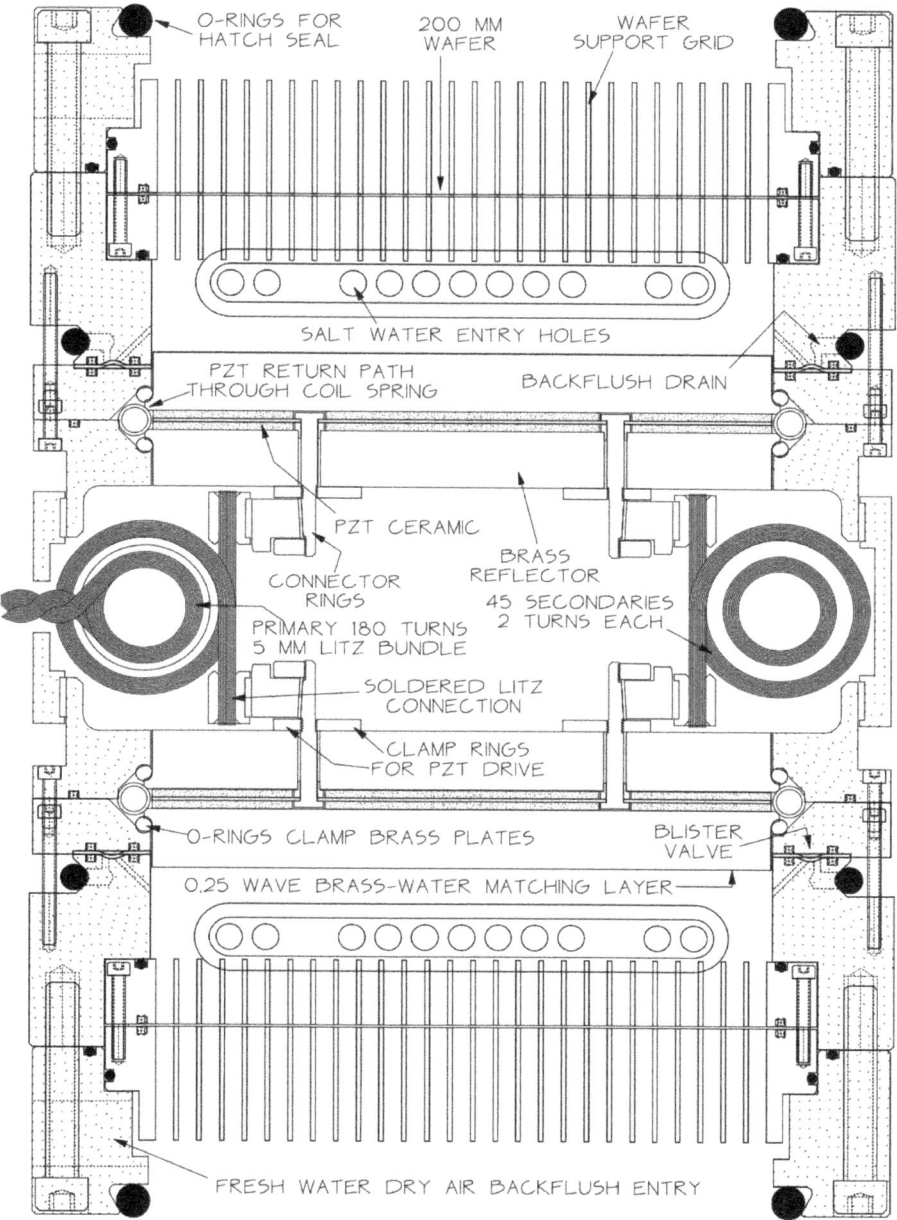

Figure 3 A cross-section through a pair of spray modules placed close back-to-back and separated by a toroidal air-cored transformer. Charge to provide the pressure pulsation flows back and forth between the two modules through the transformer's 45 secondary windings, with losses made up by the primary winding. The modules can be sealed by hatches to allow desalinated water to be pumped backwards through the wafers which are clamped between a pair of stainless steel grids. The PZT4 piezo-electric elements are exactly what would be used at frequencies of about 45 kHz in ultrasonic cleaning baths and will assist back-flushing of any clogged nozzles resulting from imperfect primary filtration.

from one to the other, through 45 bundles of Litz wire each of which forms two turns of secondary winding of an air-cored toroidal transformer. The primary of the transformer has two layers of 90 turns. With a turns ratio of 90 and with 45 secondary windings in parallel, the current cycling back and forth between the PZT ceramics is over 4500 times greater than the current in the primary winding.

4   The Wafer

Microfabrication technology allows enormous numbers of very small features to be produced by successive stages of deposition of a wide range of materials, application of photo resists and etching. The only limitation is that everything must exist in a world of only 2 dimensions. The favourite base material is silicon crystal. This is extremely strong but also extremely notch sensitive. Careless handling of what seems a robust component but which has an undetectable scratch can leave the user holding a large number of exactly rectangular and extremely sharp razor blades which cost £1000. Notch sensitivity is not a feature of silicon nitride which can be deposited on both sides of a silicon wafer. The deposition temperature is quite high and the contraction on cooling puts the silicon into a most desirable compression.

The main requirement is to produce an array of billions of holes that are as near as possible identical in both outlet diameter and cross section, where the outlet diameter is likely to be in the order of hundreds of nm. Achieving these small dimensions repeatedly and uniformly across 200 mm diameter silicon wafers poses a major challenge to both the photolithography and etching processes.

The viscous pressure drop through a nozzle depends on the inverse fourth power of the passage diameter. An interesting feature of the cubic lattice of a silicon crystal it that it can be etched to form a pyramid shape with a half angle of 35.26°. This means that we can have a low viscous pressure drop most of the way to a small exit orifice as shown in Figure 4.

The use of silicon on insulator material (SOI) reduces the variability of the process and simplifies the fabrication sequence. SOI wafers consist of a three layer sandwich: the top 'device layer' of single crystal silicon is relatively thin; then there is a buried layer of silicon dioxide; then the bulk silicon or 'handle' wafer. The small diameter nozzles will be created in the thin device layer using an etch process that terminates at the buried oxide layer. Similarly, the other side of the wafer (the 'handle' side) will be etched to create larger diameter holes with the etch process terminating at the oxide layer. Finally, the oxide layer will be removed by etching from the handle side, thus leaving an array of spray nozzles that are open all the way through the sandwich.

The detailed process sequence to achieve these features will require optimisation but, in outline, it will be as follows.

The SOI wafer will first be thermally oxidised and then a layer of silicon nitride will be deposited on both sides using a low pressure chemical vapour deposition system (LPCVD) at about 850 °C. The resultant nitride film will be

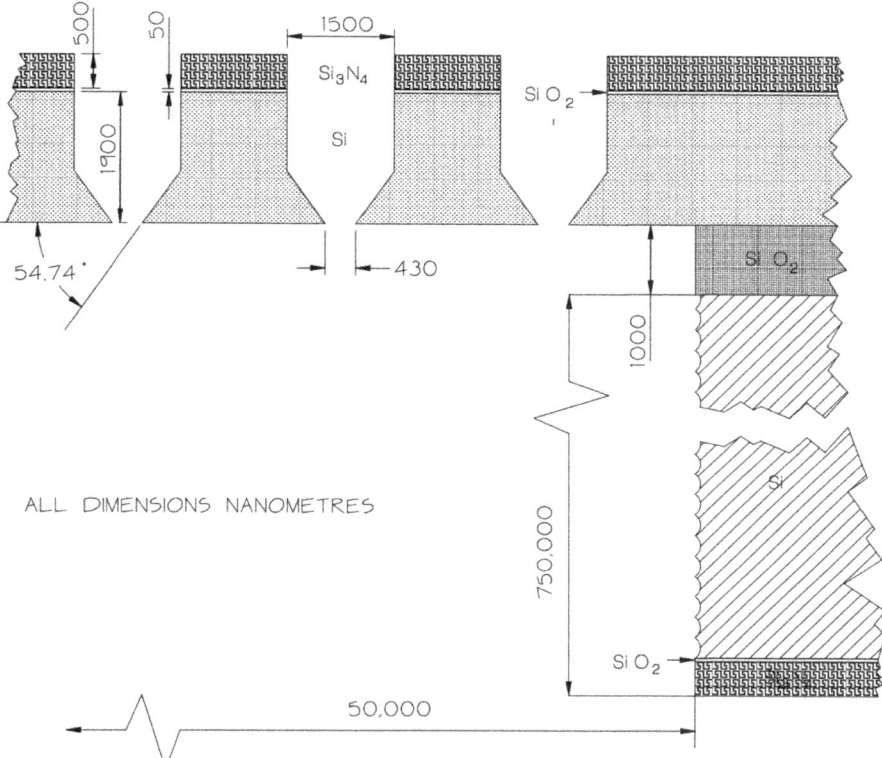

Figure 4 The proposed dimensions of nozzles in nm. A large number of 430 nm nozzles with tapered entries meeting in a 50 μm hole through the 750 μm dimension of each 200 mm wafer. The 'tide marks' in the thick silicon show successive layers of etching of the Bosch process.

in tension, thus improving the mechanical strength of the wafer, and silicon nitride is a very good masking material for the etch processes which will be used.

The outer nitride–oxide film on the device side will be patterned using photolithography and etched to create windows of the appropriate dimension. Then, a wet etch process using TMAH or KOH giving an etch rate that is sensitive to the crystal orientation will form pits in the thin silicon device layer with sidewalls at 54.74° to the surface. The etching will stop when the oxide layer is reached.

Variations of etch rate across the 200 mm diameter wafer have to be taken into account as these can result it variations in nozzle diameter. It is proposed to form the cylindrical part of the nozzles in a relatively thick layer of silicon nitride using reactive ion etching (RIE) and then form the tapered section in the silicon of the thin device layer using wet etch. The use of a thin device layer combined with an etching where the sidewall angles are defined by the crystal planes helps to reduce variability. An indicative arrangement is shown in Figure 4 with some tentative dimensions – these are likely to change as a result of process optimisation.

Again, on the handle side of the wafer, the outer nitride–oxide film will be patterned to form windows of the appropriate size, around 50 μm. The handle wafer will be around 730 μm thick so a deep silicon reactive ion etch (DRIE) based on the Bosch process will be used to form holes with almost vertical sidewalls, all the way through the wafer, stopping at the oxide layer. These deep silicon etch processes typically use SF_6 as the source of etchant (F) and C_4F_8 as the source of passivation (polymer) material to protect the sidewalls.

After inspection to confirm that all the silicon has been etched from both the small nozzles and the larger holes, the oxide layer will be removed by reactive ion etching from the handle side.

The attenuation of ultrasound at tens of MHz is very high. The ideal solution would be to include the piezo-electric excitation in the body of the wafer or its support grid. At present this looks very difficult but microfabrication engineers have a long track record of achieving the apparently impossible in a few Moore's law cycles.

5 Filtration

If spray generation was the most challenging part of the project, the second is filtration to the level needed to prevent nozzle clogging. Before the use of the Salk vaccine, polio caused many deaths and many more cases of permanent paralysis. Ultra-filtration technology was developed to remove the 29 nm diameter polio virus from drinking water. This technology is now used in very large quantities for pre-filtration of sea water going to reverse osmosis membranes.[11] Filters clean up feed water taken from quite near the coast which will be a more severe requirement than with our mid-ocean water. The suppliers of Seaguard X-flow filters, the Dutch company Pentair, will guarantee operation for 5 years and expect most to last for 10. The present installed capacity is $6\,000\,000\,\mathrm{m^3\,day^{-1}}$ and plant with a capacity of a further $3\,000\,000\,\mathrm{m^3\,day^{-1}}$ is under construction. These volumes are considerably larger than inkjet printer consumption. While filter engineers are confident about filtration to the level required they may have been doubtful about the feasibility of inkjet printing!

The plan is to run a set of eight filters in parallel with some of the water from seven filters going to back flush number eight and to change the back flushed one every few minutes. The block diagram of the system the filtration system is shown in Figure 5.

Each filter module needs two valves working in sea water and the output from each bank of filters must be sent either to a spray head or to the other filters. Valves must control the flow of salt water and, after the filtration stage, must not produce any wear debris which could easily be produced by the sliding motion of a spool valve. The proposal is to use blister valves. These are the watery equivalent of the field effect transistor. Large numbers can be formed by a single sheet of rubber clamped between two plates of ABS plastic. Figure 6 shows a section of a blister valve.

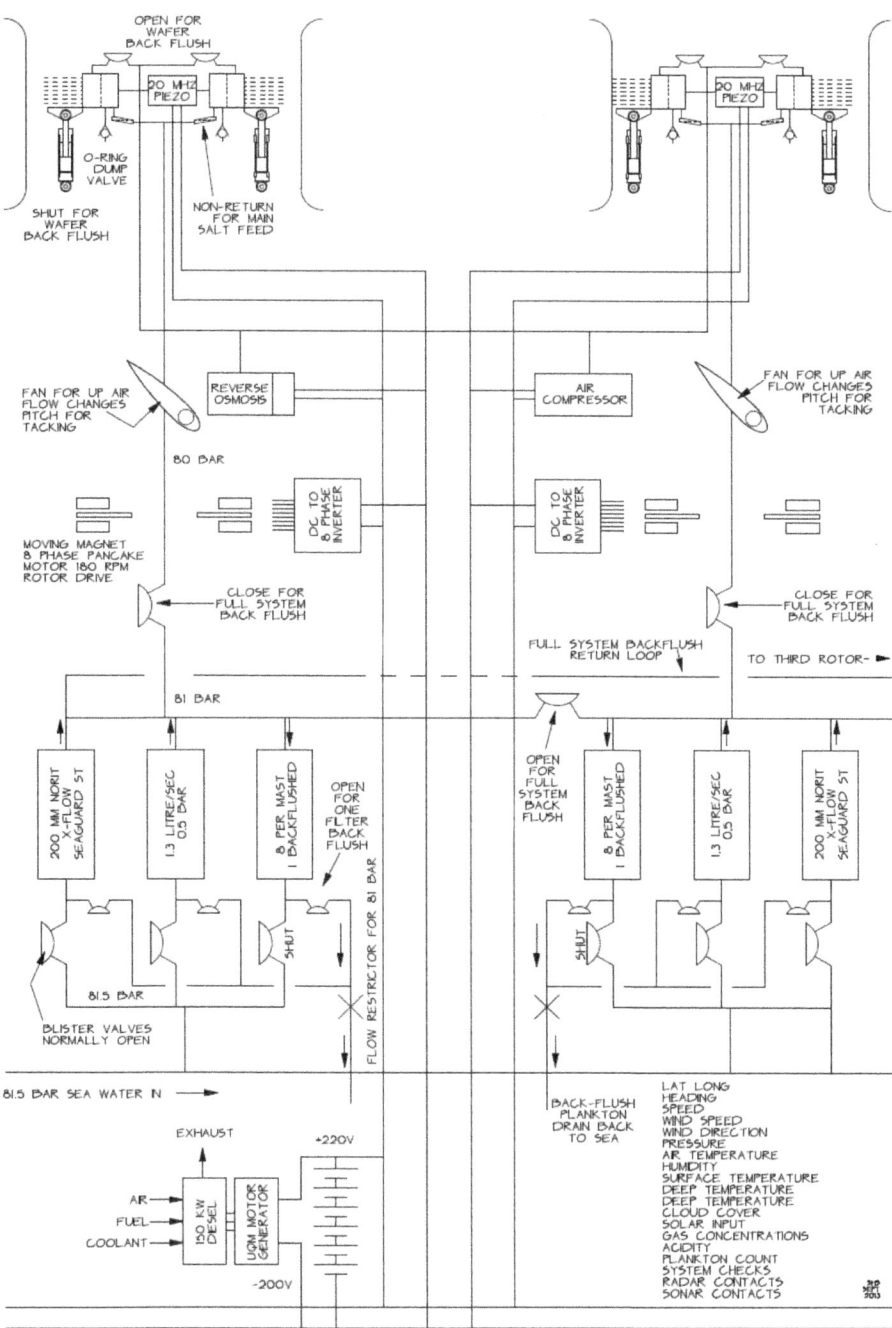

Figure 5 Plankton are removed by banks of 8 filters (3 shown) with sequential back flushing.

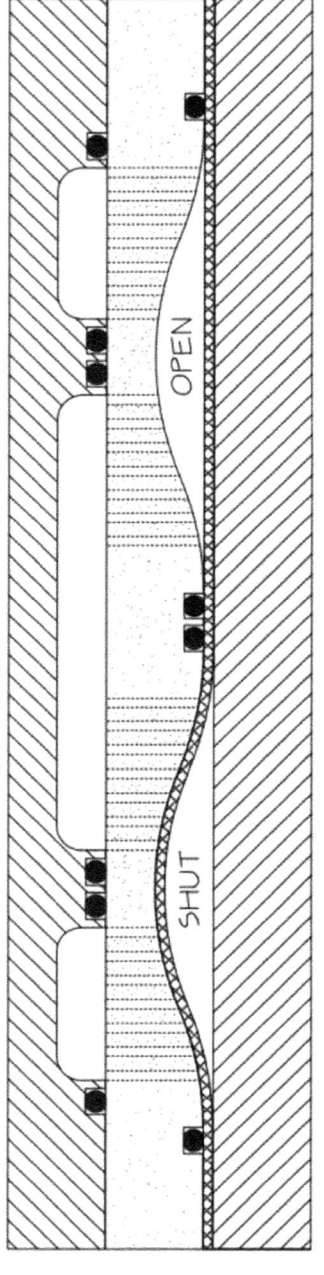

Figure 6 A sectional view through a blister valve. Oil pressure on one side of an elastomeric membrane can block a grid of passages on the other side with the minimum creation of wear debris. The similarity to a field effect transistor is not accidental. Two valves for each of eight filter modules plus two more for back-flush connections between filter sets can be formed by a single rubber sheet. It was frustrating to learn that the idea and even the name 'blister' had been anticipated by Perkin Elmer in 1996.

Flow into the valve comes through a large number of small holes in the upper block, through a smooth-edged cavity and out through a second set of small holes. If hydraulic oil at a pressure slightly greater than the pressure in the salt water is pumped into the space below the cavity, the rubber will rise to the ceiling of the cavity and stop the flow. The oil flow for each blister valve will be controlled by an electromagnetically operated poppet valve.

In order to perfect the operation of the Pentair Seaguard filters, however, we have to expect more contamination from the pipe work connecting the filters to the spray heads. We have to transfer the cleanliness standards of a micro-fabrication clean room to a ship yard. It is a sobering thought that a spherical blob of tar in cigarette smoke is perfectly sized to clog a spray nozzle. Furthermore we must not allow salt water to dry out or fresh water to freeze in the nozzles. It will therefore be necessary to back flush the silicon wafers, with fresh water followed by super-filtered dry air.

The wafer back-flushing needs the provision of a hatch to close the exit of the wafer housing. The hatch will be closed, the salt feed cut off and fresh water pumped backwards through the wafer nozzles. We can now use the high frequency ultrasonic system at much lower frequencies as an ultrasonic cleaner which it so closely resembles. Figure 7 shows a plan view of the spray head.

Most of the dirt in hydraulic systems is built in from the start. The first step in the clean-up will be to make all fittings leak proof and pump the system to a high pressure, 50 to 100 bar, with super filtered air. This will be released abruptly through a large steel ball retained in a cone by a magnet flux. The sudden release of pressure will cause sonic pressure pulses along the pipes to remove debris. The process will continue until the air coming out is deemed sufficiently clean.

The system will be resealed and pulled down to a hard vacuum. Then a heater will be turned on in a basket of a material known as Parylene. This converts it into a vapour which instantly spreads to the entire inner surface area of the system. Gas molecules get everywhere and condense to form a uniform layer over the remaining dirt on pipe walls.

6   Vessel Design

The places where marine cloud brightening will be most effective are in clean air far from land, but these locations vary with the seasons and so hardware should be mobile. Supplying conventional fuel, food, water and medical attention to mid oceans is expensive. This suggests that spray should be produced by unmanned, wind-driven vessels perpetually cruising the oceans and dragging some form of generating plant through the water to provide energy for the spray equipment. Initial ideas for engineering hardware are given in a paper by Salter *et al.*[12] Instead of textile sails the driving force would come from Flettner rotors. These are vertical spinning cylinders which can produce much higher force per unit area than sails.[13] The speed of rotation acts in a way similar to the angle of incidence of an aerofoil but is

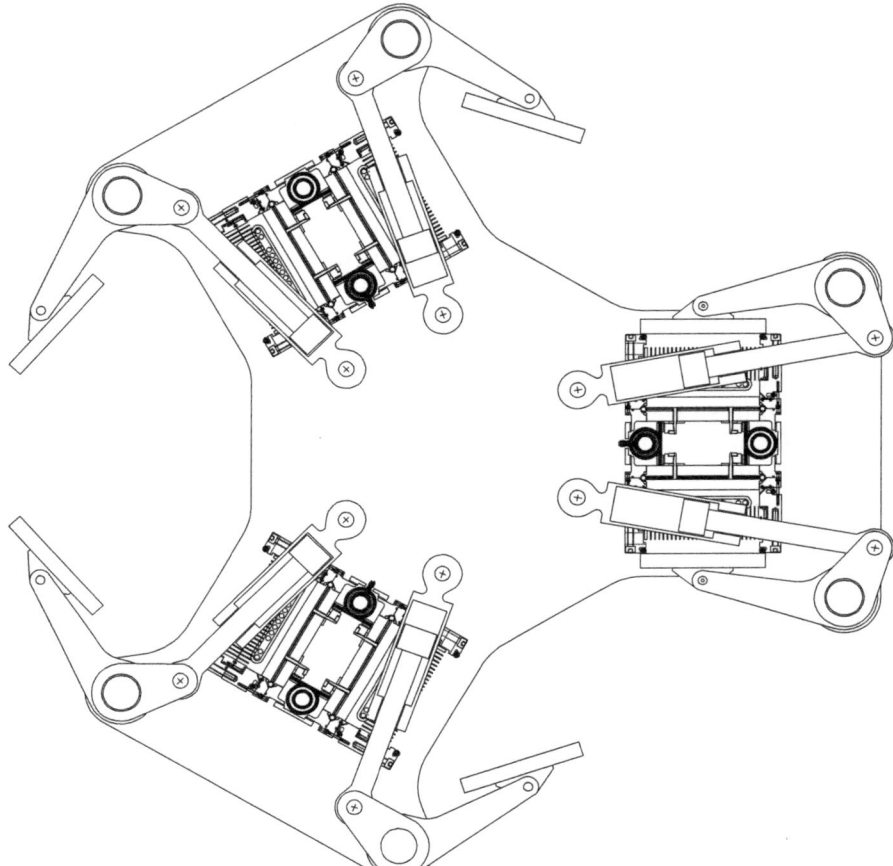

Figure 7 The spray head seen from above, showing open and shut wafer hatches. Shutting the hatch allows the wafer nozzles to be back flushed with de-salinated water and dried with ultra-filtered air. Putting wafers back-to-back shortens the distance over which the large charge which actuates the piezo-electric elements has to travel. The exit passages are horizontal so that any water can be cleared by pitch and roll motions. A fan below the spray head will move a vertical air stream at 9 m s^{-1} towards the reader.

much easier to control by computer. The high agility of Flettner rotors relative to textile sails is a particular attraction. Rotor-driven vessels can stop, go directly astern and rotate 180° in either direction about their own centre or any other point.

Spray will be blown up through the rotor by an air flow of 9 m s^{-1}, just less than the Weber coalescence velocity and will emerge at a height of 25 m.[14] It will be entrained with surrounding air and the mixture will rise to a height of about 5 rotor diameters above the rotor top. The relative humidity very close to the water surface will be nearly 100%. This falls to typically 60% at a few metres and then rises slowly to 100% at the cloud base. The very large surface area of the spray means that evaporation of the spray plume up to 100% relative humidity will be fast, leaving large numbers of liquid drops. The upward velocity will fall rapidly above 35 m. For drop sizes of 800 nm, we have to increase the Stokes prediction for the still air falling velocity from 19.8 to 23.8 $\mu m \ s^{-1}$.[15] However, this is very small compared with the turbulent velocities in the marine boundary layer which is an appreciable fraction of the local wind speed but with a vertical component clipped at the surface.

Latent heat for the first stage of evaporation will come from the surrounding air of the expanding plume. We can use psychrometric charts (a graphic representation of the physical and thermodynamic properties of airwater mixtures) to get the temperature drop. For an input humidity of 60% and a dry bulb temperature of 15 °C, the temperature drop in the wake will be about 3 K. The resulting density increase will mean that the cooled plume will fall rapidly and spread out over the sea surface taking heat from the water below for any further evaporation. The air above the plume will cool at night but the water below will stay at almost the same temperature causing most of the subsequent rise.

The tops of the original Flettner rotors were fitted with a flat disk to reduce air flow from the high to the low pressure side of the rotor. Alexander Thom argued that the addition of extra disks would increase the lift coefficients and reduce drag coefficients.[16] This was supported in practical tests by Norwood.[17] More recent computational work by Craft *et al.* has shown that the lift coefficients of 10 and above predicted by Mittal and Kumar at low Reynolds numbers,[18] also applied to ones at 10^6.[19] They also found that drag coefficients fell with increasing spin ratios down even as far as the negative ones suggested by Mittal and Kumar at low Reynolds numbers. They found that the increased lift coefficients at high spin ratios due to extra fences were not as high as Thom had hoped and that fences needed higher drive torque. However, most importantly, they found that drive torque coefficients of 0.0025 were encouragingly lower than those predicted by Glauert.[20] Unfortunately the drive torque coefficients of the 10 500 tonne Enercon Flettner ship have not been published.

Mittal and Kumar showed that there were critical rotation speeds which produced large oscillatory forces from vortex shedding. At a spin ratio of 4.5, with a predicted a lift coefficient oscillating between 21 and 23, the instantaneous drag coefficient ranged from +0.7 down to −1.5. Forces were stable between spin ratios between 2 and 4.3 and again above 4.8. However, these oscillations were not reported for either of the Flettner ships and were not observed in the trials of Cloudier, a 37-foot sea runner converted to rotor propulsion by Marples for a television production company. It may be because the computer modellers use the numerical equivalent of rigid mountings for their rotors. A partly flexible rotor on a more flexible mast on a ship which is free to roll to make waves will provide damping tending attenuate any oscillatory forces.

The first rotors used by Flettner for his 1926 Atlantic crossing were made of steel but, even so, weighed only one quarter of the rigging they replaced. With modern composites they can be made even lighter than a conventional mast because a thin-walled tube has such a high structural efficiency. Indeed the wall thickness needed to resist the rotor bending stress in a Category 4 hurricane is so low (0.7 mm) that the failure mode would be buckling. The rotors will therefore have a double skin separated with corrugations. With foam filling they can provide buoyancy to prevent total capsize. The structural design is driven not by stress but by the need to keep deflections within the angular range allowed by a pair of SKF spherical thrust bearings. The high lift coefficients mean that a given thrust can be achieved with a lower centre of pressure so that heeling moments on the hull and bending stress at the mast foot are reduced. Indeed when the wind is from the quarter a rotor vessel heels *into* it.

We require the spray generation system to be above the spherical bearings which support the weight of the rotor and transmit its wind loads. This means, however, that salt water, fresh water, dry air, electrical power for piezo-excitation, high pressure oil for hatch operation, medium pressure oil for blister valve operation and oil return line have to pass through the 150 mm bores of a pair of SKF 29330 bearings. The SKF data shows that we could lift the entire vessel out of the water and spin it on one bearing.[21]

7 Justification of the Trimaran Configuration

Let us recall the evolution of ship forms. The original mono-hull form remains the favourite for heavy cargoes and large numbers of passengers. Internal spaces are large and have convenient shapes. Most of the hull surfaces are easily formed from single curvature plate. Roll may be high but recovery from roll is usually certain. The wave-making resistance rises very sharply at speeds approaching a value (in m s^{-1}) of 1.34 times the square root of the water line length (in m) and many mono-hulls never exceed this. There is therefore a strong incentive to build long vessels.

The wave-making resistance of a catamaran is lower than that of a mono-hull and shows smaller humps in the wave-making drag curve, especially if the demi-hulls are staggered so that the waves of one hull interfere with those of the other. Surface shapes are more complicated. Internal space is less convenient. Roll is less than that of a mono-hull but, if capsize does occur, recovery can be problematic. Capsize stern over bow, known as 'pitchpoling', can sometimes occur when vessels are driven too hard. Waves give catamarans more uncomfortable, or perhaps a more exciting, ride with appeal being a factor depending on the age and courage of the skippers.

Trimaran enthusiasts honour the Polynesian inventors by retaining their words. The central 'vaka' (the main hull) is joined by one or more 'akas' (the supports) to two 'amas' (the outriggers). The trimaran configuration is now the preferred configuration for very high performance vessels needing both speed and survival in rough, round-the-world conditions. The vaka can have adequate internal space for spray equipment. Roll resistance and problems of recovery from capsize are similar to those of the catamaran but comfort in oblique seas is better. The correct staggering of the amas can produce a very high degree of wave cancellation for particular chosen speeds but for cancellation at the very highest speed, the amas have to be placed so far forward, or so far aft of the vaka, that structural integrity is questionable.

Trimarans (and also other vessels) can be capsized by very steep waves which can occur by rare phase combinations of the components of the wave spectrum, from refraction over shallow water such as the Labadie bank between the Scillies and Fastnet or when waves meet an opposing current such as the Aghulas off the east coast of South Africa. The probability of a capsize can be reduced by careful course planning and spray vessels are more likely to operate in the summer hemisphere. The use of unmanned vessels changes the safety argument. If the propulsion force for a vessel is a strong driver of its cost, we should calculate how the extra cost multiplier of monohull drag compares with the fraction of trimarans which might be lost in extreme conditions.

The design of a trimaran geoengineering spray vessel has been inspired by record breaking yachts such as Banque Populaire V and Hydroptere.[22] However, it differs by the need for a heavier payload due to spray generating plant and the need to generate quite large amounts of power – perhaps 300 kW – for spray production. Banque Populaire displaces 23 tonnes and Hydroptere only 7.5 tonnes with only 0.9 m^2 of living space per crew member. A spray vessel will have the same 40 m waterline length as Banque Populaire V but, with the additional spray equipment, may displace as much as 90 tonnes. It would of course be quite improper for decisions about vessel design to be influenced in the slightest way by the prospect of making a contribution to record-breaking yacht design.

The 2008 spray vessel design had two ducted turbines, very much larger than any propeller for that size of vessel. The ducts round the turbine rotors would reduce tip vortex losses and could provide the function of a keel to resist beam forces. Generation could be done with rotating permanent magnets in a rim generator with flux cutting speed limited by cavitation. Alternatively, a more conventional generator could be driven at higher speeds through epicyclic gearing. The blade pitch and chord taper are chosen to give an equal pressure drop along the span.

One unfortunate feature of these arrangements is that cavitation problems are unevenly shared, being much worse at the blade tips. Another issue could be the possible requirement for higher speed during the non-spray mode, to move fleets of spray vessels for tactical spraying at sites in the opposite hemisphere. It also turns out that, even with good hull designs, the force needed to drive the hull is much larger than the force needed for the turbine. Operating a turbine over a wide range of speeds seems to require variable-pitch blades without much room for the pitch change mechanism. These thoughts have led to a new proposal for four, separately flapping akas hinged at the junction of the hull with high pressure oil hydraulics to control the amas for roll stabilisation. The concept draws heavily on ideas for vehicle suspension, first used by Citroën in 1952 and later adopted by several other motor manufacturers,[23] and combines them with the hydrofoils of Hydroptere.

In calm or very low wind speeds the vessel would be supported by the buoyancy of the vaka and amas. However, hanging down below each ama will be a submerged hydrofoil with variable pitch as shown in Figures 8 and 9. When the wind increases enough for the forward speed to reach a critical value for take-off, probably 4 m s^{-1}, the pitch of the hydrofoils would

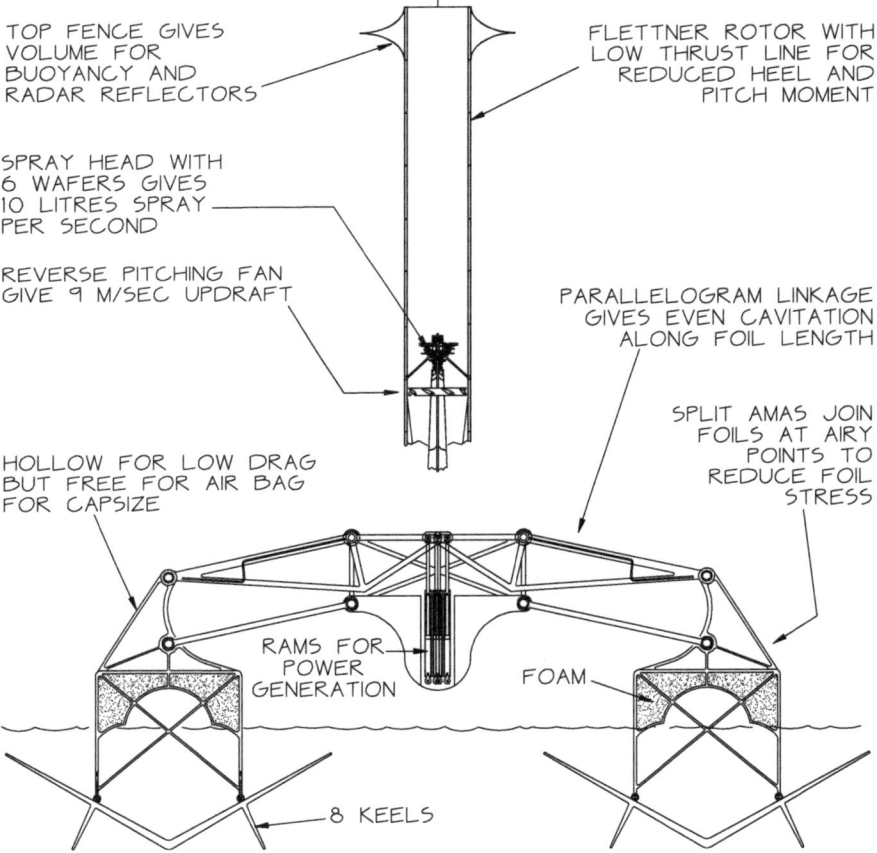

TOP FENCE GIVES
VOLUME FOR
BUOYANCY AND
RADAR REFLECTORS

FLETTNER ROTOR WITH
LOW THRUST LINE FOR
REDUCED HEEL AND
PITCH MOMENT

SPRAY HEAD WITH
6 WAFERS GIVES
10 LITRES SPRAY
PER SECOND

REVERSE PITCHING FAN
GIVE 9 M/SEC UPDRAFT

PARALLELOGRAM LINKAGE
GIVES EVEN CAVITATION
ALONG FOIL LENGTH

SPLIT AMAS JOIN
FOILS AT AIRY
POINTS TO
REDUCE FOIL
STRESS

HOLLOW FOR LOW DRAG
BUT FREE FOR AIR BAG
FOR CAPSIZE

RAMS FOR
POWER
GENERATION

FOAM

8 KEELS

Figure 8   A front view of the trimaran spray vessel with hydrofoils below four amas. The mean load on each foil will be one quarter of the vessel weight. However, alternating variations of the hydrofoil pitch angle can be used to control roll, pitch, heel and trim and will generate large amounts of power on board the vaka to generate spray and spin the rotors. The dihedral angle of the hydrofoils means that at high speed the wetted area can be very low but the transfer from ama buoyancy to foil lift can happen at low velocities. Most of the energy taken from the wind will go into spray production rather than overcoming vessel drag. Movement of the amas by swell in windless conditions can provide energy for communications and controls. Hydrofoils can also be used for propulsion.

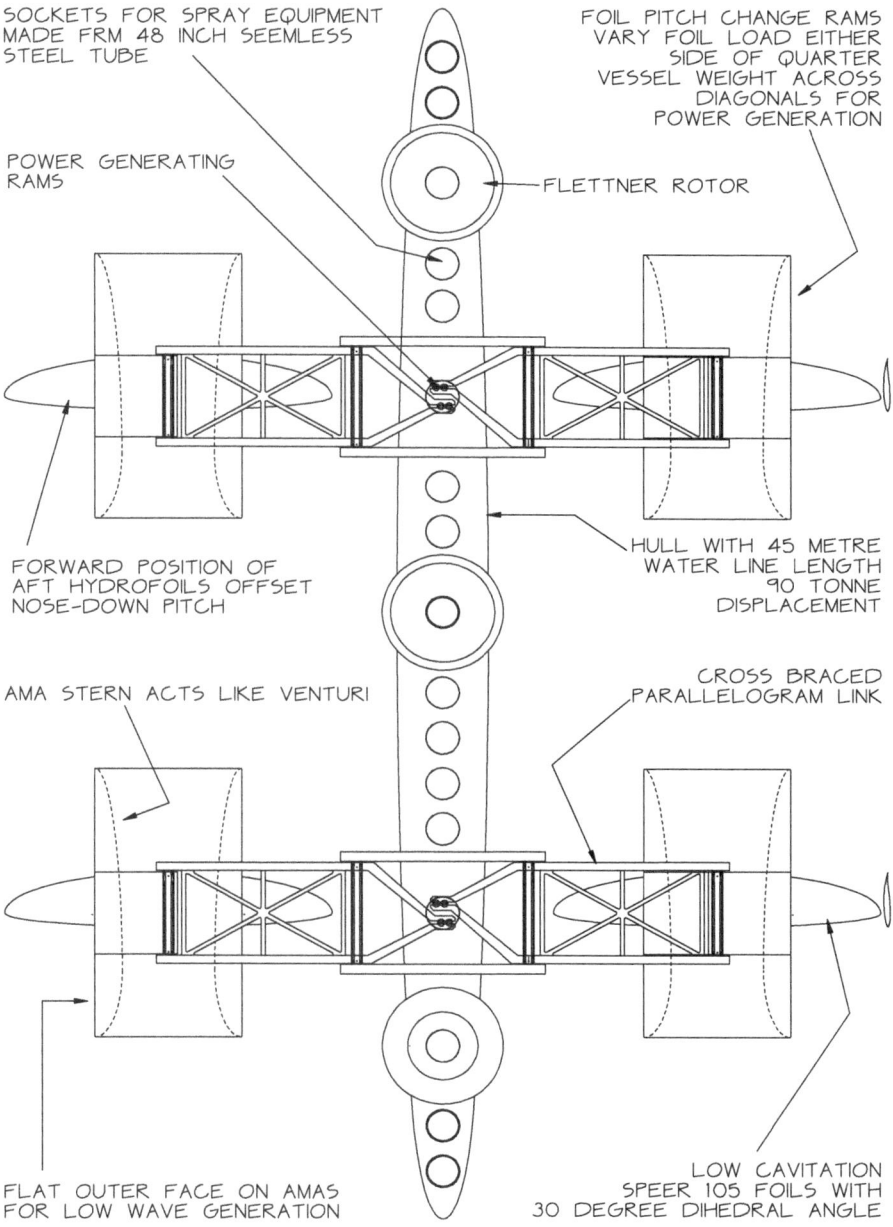

Figure 9 A plan view of the spray vessel. The four amas have been split into two sections so that hydrofoils can be mounted at their Airy points (the points used for precision measurement that support a bar horizontally to ensure minimal bending). This reduces foil stress by a factor of 5 relative to support at two ends and by a factor of 20 relative to a cantilever support. The flat outer faces of the amas will give a low wave-making drag. The gradual expansion of the inner section will act like the exit passage of a Venturi (a short piece of narrow tube between wider sections for measuring flow rate or exerting suction).

be increased so that each could support a quarter of the vessel weight. With fast control of the hydrofoil pitch angle we have the ability to control roll, pitch and heave motions. The foils are supported by a pair of vertical struts with a separation chosen to equalise the hogging (curving upwards) and sagging (curving downwards) stresses analogous to Airy points. Buckling of the struts will be critical and so we will use carbon fibre. Many hydrofoil craft use quite large dihedral angles. At higher speeds the vessel rises in the water thereby reducing the wetted area of the foils and there is an automatic control of roll and pitch. The transition to foil-born mode can take place at low speeds. However, large dihedral angles make the problem of pitch variation more difficult because the pitching moment on the foil will vary with depth of immersion. If we have confidence in the control loop for foil pitch it would be possible to have fully immersed foils with no dihedral angle. This would remove some of the anxiety about droplet erosion from spray at the surface but would lead to an increase in wetted area which, at the moment, seems more important.

Rigidly connected amas and hydrofoils will suffer severe wave loading at high speed in rough seas. Accelerations for a given bump amplitude rise with the square of encounter velocity, and this sets an upper limit to operating speed. Allowing controlled movement of the amas is like fitting independent suspension and shock absorbers to an unsprung cart, and can push the operating envelope to higher speeds and rougher seas. Furthermore, the damping of a shock absorber can be replaced by a mechanism which can generate energy.

Suppose that instead of equal sharing of the vessel weight between four hydrofoils we momentarily increase the pitch angle of the fore port foil by one third and decrease it for the fore starboard. This would produce a roll torque on the fore section which can be exactly balanced by a reduction in foil pitch on the port aft and an increase at the starboard aft.

When the akas approach the end of the allowable travel we reverse all four pitch angles. This would induce large alternating torsional stresses in the mid-section of the vessel but would not affect pitch or roll. The resulting movements of the two upward moving foils with the large forces would do work on hydraulic rams. The movement of the akas would resemble that of some water walking insects and might appear to be an ungainly waddle but the vaka will advance smoothly and steadily. Four akas moving through 1 m against a third of the weight force of 90 tonnes will produce >1 MJ per stroke. While conventional turbines are designed to maximise energy production per unit of swept area with less consideration of the thrust, the opposite is the case for this application. We can afford to sweep a large area but want to minimise drag.

Foil angle adjustment is done by tilting the foil support struts about an axis through the ama with a pair of hydraulic rams either side of the ama above water level. The SKF company offer spherical plain bearings with amazingly low friction coefficients of 0.025. However, the entire weight of the vessels plus a factor for power generation must be taken by just eight bearings. Their life prediction equations indicate 4 years of operation. The angular motions are only a few degrees and so are uncomfortably small for rolling bearings. The proposal is to buy SKF parts and use spark erosion to cut the pockets for hydrostatic bearing pads fed by oil pressure from the power take off rams. The small angular deflections mean that we can seal everything with corrugated rubber gaiters and recycle the oil flowing out of the bearings. The obvious new design problems are extra moving parts, the conversion of irregular reciprocating forces to electricity or steady pressure oil flows, stress reversal leading to fatigue, multiplication of the weight force of the vessel by the leverage ratio, passing these forces through bearings, cavitation and the need for rapid pitch control.

If these problems can be solved there may be a number of advantages:

- Transient loads from wave impact can be reduced.
- Roll, heave and pitch motions on the vaka can be controlled.
- Heel and trim can be controlled to suit wind direction.
- Most of the power generation mechanisms can be placed inboard above the surface.
- Flotsam can be avoided.
- Amas can be lifted out of the water for inspection and maintenance.
- Cavitation pressures can be made the same along the span of the hydrofoils.
- The sweep velocity can be set according to power requirements and vessel speed.
- Driving the amas as wave makers can produce a side thrust.
- Driving the hydrofoils with the right foil pitch variation can produce propulsion.
- The pressure in hydraulic rams can be used to feed oil to hydrostatic bearings.
- It may be possible for ama movements and air bags to aid recovery from capsize.
- Moderate amounts of power, enough to preserve communications and essential control functions, can be generated from wave motion with no local wind.

Several of these requirements can be made easier by the use of newly developed digital hydraulics technology which originated in the needs of wave energy.

8   Digital Hydraulics

Work on power conversion for wave energy has showed that conventional hydraulic machines did not have suitable power ratings, flexibility of control and part-load efficiency for the demanding requirements of waves, especially the wide range of amplitudes. Conventional fast machines use an axial configuration for the chambers about the rotating shaft and the variation of the angle of a swash plate or axis of a cylinder bank to change the volumes of oil delivered per rotation. This can take seconds on a large machine so the bandwidth and loop gain of feedback loops are low. There is one inlet and one outlet port for each shaft and so it is difficult to control oil flows to and from different sources and sinks. Quite large surface areas have to move at high speed very close to one another so that there is an awkward compromise between shear and leakage losses. Pressure and the resulting leakage are present all the time even if no work is being done. The fine moving clearances must not be affected by the large forces.

The new fast digital hydraulic machines use a radial configuration so that one crank shaft can drive or be driven by pistons in many banks.[24,25] The geometrical changes of swash plate angle are replaced with decisions about the state of two electromagnetically controlled poppet valves at each chamber, one going to an oil tank at a low boost pressure and one going to a high pressure gallery. Poppet valves have a higher dirt tolerance than the sliding surfaces at a conventional port face. With clean oil they have virtually zero leakage. Unlike port faces and spool valves they can continue to function despite seat wear.

The shaft will usually run at a steady speed, often 1500 or 1800 rpm. Decisions about the valve state are taken by a micro-computer at times close to top or bottom dead centre for that chamber. This allows each chamber to idle, to pump or to motor for the next half rotation of the shaft but never requires a valve coil to oppose the force of high pressure oil. With six chambers in four banks there will be 24 decisions made for each shaft rotation so response can begin in 2 ms and get to its full magnitude in half a rotation *i.e.* 20 ms. Variable-rate pumping is achieved by the decision whether or not to hold open a low pressure valve at bottom dead centre. If the valve is held open, oil will flow back to the tank. No work will be done. The energy in flow losses will be about 1/500 of the work that would have been done if the valve had been allowed to close and oil delivered through the high pressure valve. If motoring action is required the high pressure valve will be held open at top dead centre so that oil can flow in to the chamber and do work on the crank. Just before the shaft reaches bottom dead centre, however, the high pressure valve is closed and the very last part of the stroke is done with energy stored in the finite bulk modulus of oil and elasticity of the mechanical parts. The timing is chosen so that chamber pressure is the same as the low pressure tank at bottom dead centre so that the low pressure valve can open to allow the oil to exit. The high pressure valve will now be very firmly closed against its seating by the high pressure. The piston will rise, delivering oil to the tank, but just before it reaches top dead centre the low pressure valve will close and oil in the chamber will be compressed up to the high pressure needed to open the high pressure valve, so that the cycle can begin again. Valve operation always takes place when there is a low pressure difference across a valve and so the machines are quiet, making less noise than the induction machine which drives them.

Nearly all the forces inside a digital hydraulic machine are due to an oil pressure. This means that the pressure can be fed to the far side and a part under load to form a hydrostatic bearing with either very low or zero contact force. If, a rather big if, oil can be kept clean, then the life of hydrostatic bearings can be made indefinitely long.

A further application of digital hydraulics in a spray vessel is the generation of the pressure for required for spraying. We first thought that the pressure needed would be suitable for down hole pumps such as the Grundfos SP series. These combine a long stack of hydrokinetic impellers with a three-phase ac motor in a package which will fit in a drilled well-shaft. Stainless versions can operate in salt water. Efficiencies of 80% at the lower pressures are possible but these efficiencies are reduced at the higher values of pressure which have been shown to be necessary by the COMSOL Multiphysics analysis, which showed that 80 bar would be needed for monodisperse spray at 800 nm.

Digital hydraulics favours high pressures but low flow rates. While there has been a steady increase in pressures used for hydraulic machines, there is a present upper limit of 400 bar set by the availability of commercial seals and fittings. However, it is possible to design pressure exchangers in which a cylinder is fitted with two linked pistons of differing diameter. These will enable the step down from 350 bar to the 80 bar suitable for nozzles, with a corresponding increase in flow volume. The 80 bar oil must be separated from water. The rubber separating membranes used for gas accumulators are notoriously unreliable, perhaps because there is no accurate control or limitation of strain. It may be possible to achieve a satisfactory fatigue life with the design shown in cross-section in Figure 10.

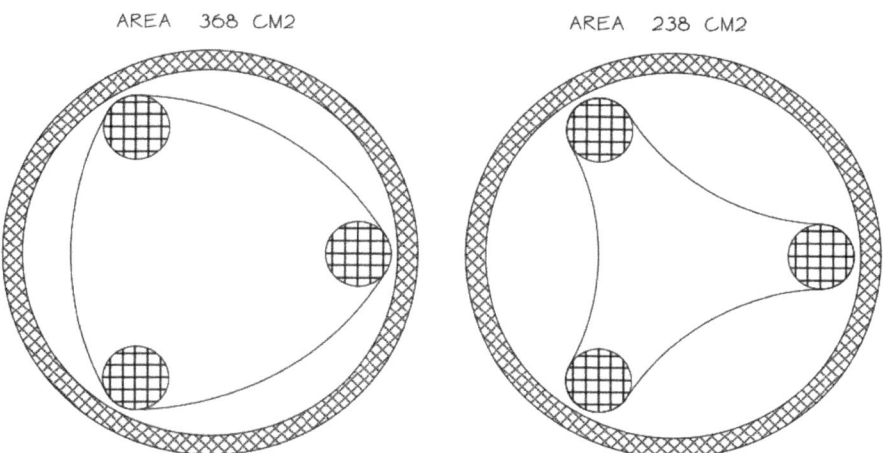

Figure 10 A sectional view of the oil to water pressure exchanger. The design is based on Cadwell's work on the fatigue life of rubber.[25] Reversal is bad. The strain in the rubber tube is only 10%.

The fatigue life of the design depends on the rubber fatigue life measurements made by Cadwell in 1940.[26] During an elaborate series of tests he showed that it was the *reversal* of strain which shortens life. If the strain goes from 100 to 125% of the unstressed dimensions his specimens were lasting nearly 10^9 cycles, while going from zero strain to 25% would give a life of only 10^7 cycles.

The body of the pressure exchanger is an open-ended tube with end caps retained by three stainless steel rods. Around the rods is wrapped an open length of rubber tubing. The rods are pushed apart by a series of increasingly larger trefoil-shaped section so that they can be forced into holes in the end caps. A final parallel set of expanders can be left in place to reduce bending stresses in the rods. Each end of the volume is closed by a fairing resembling the trousers of a fat, three-legged person. With the proportions drawn, the strain variation in the rubber is only 3%, a tenth of the strain which would fail at 10^9 cycles.

Salt water goes into the centre of the rod group through a silicon nitride non-return valve and is pumped out through another non-return valve at the other end. Oil can be put in and taken out through ports at the centre.

Figure 11 shows the block diagram of power generation for the moving akas and the pumping system from oil to water.

9 The Mathematics

So far, people modelling geoengineering have applied fixed concentrations of aerosol to selected parts of the atmosphere regardless of its state or the season of the year. For the very long lifetime of stratospheric sulfur this is a reasonable simplification. However, for tropospheric sea salt this would be similar to buying a car with all the movements ever needed for its steering wheel stored in a read-only memory. It would be odd if the phase of the El Niño oscillation or the timing of a monsoon had no effect.

One of the earliest predictions of the effects of marine cloud brightening was by Jones *et al.*[27] They tested the effects of continuous spray at three small areas, one off California, one off Peru and one off Namibia. The area totalled 3.3% of the oceans but produced a cooling of nearly 1 W m^{-2}, about two thirds of the anthropogenic thermal change since pre-industrial times. Their results predicted that spray would increase precipitation by useful amounts in several dry regions of the world but would reduce it by 0.8 to 1 mm day^{-1} in the Amazon. This is small compared with present precipitation values of 6.5 mm a day and was not shown in modelling by Rasch over a wider area.[28] However, interference with precipitation anywhere must be a cause for anxiety.

The main effect of spray is to reduce sea surface temperatures by amounts which are small compared with the change from cloud to clear sky, and even smaller compared with day to night. Lower sea surface temperatures mean less evaporation but also higher temperature gradients to produce higher wind speeds and faster water transport in monsoon winds. Rain needs large

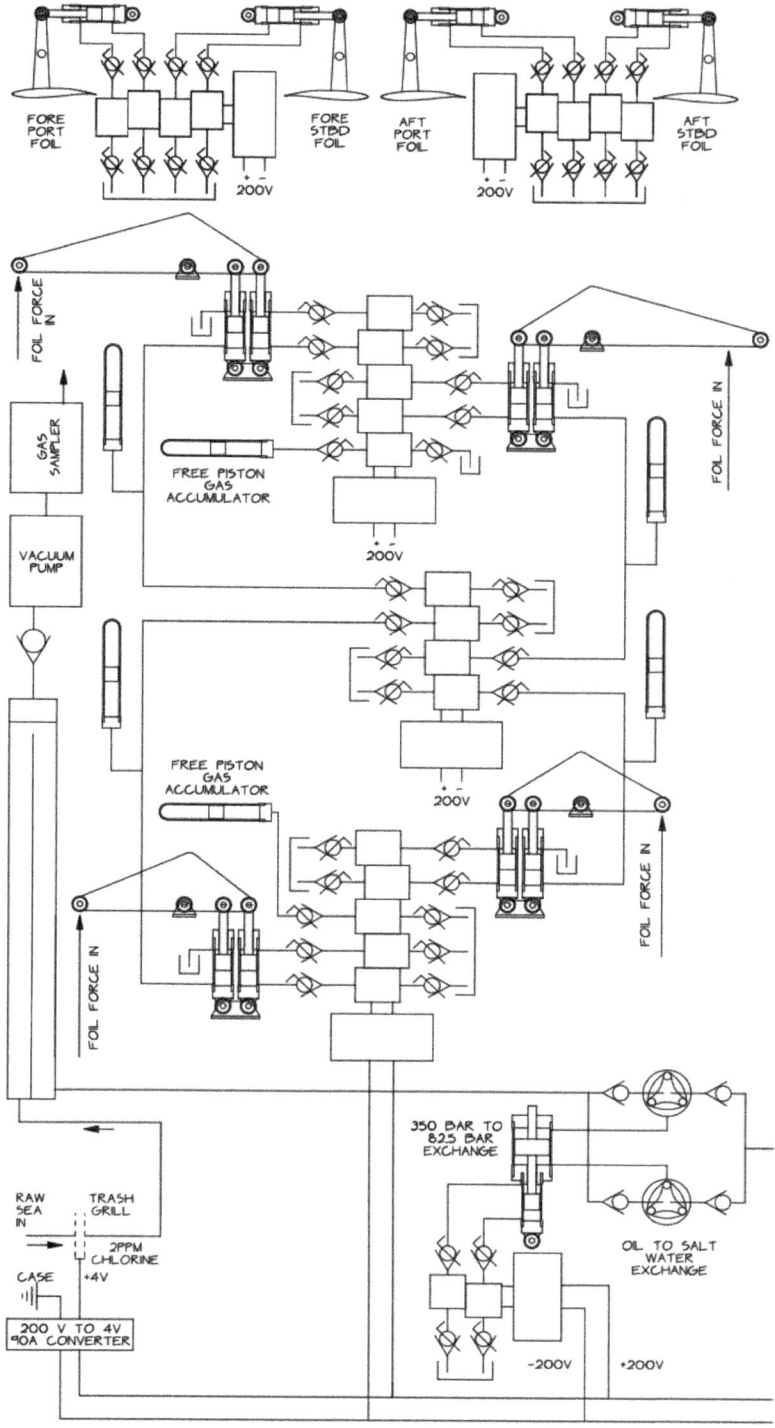

Figure 11 Multi-bank digital hydraulics are used for power generation from the moving akas and the pressure conversion from oil to water.

drops and so reducing the drop size in clouds will reduce the probability of rain but this effect will be stronger over the sea than the land and so there will be more rainfall further inland. There are at least six other conflicting climate mechanisms involved. Bala and Caldeira showed that widespread marine cloud brightening led to a small *increase* in river run-off over the Indian sub-continent.[29]

We may be able to employ a technique from electronic signal processing to use climate models to produce an everywhere-to-everywhere seasonal transfer function of all the side effects of cloud brightening. The first experiment on the idea started with a real day-by-day temperature record over 20 years. The annual variation and a smaller component with a six month period were subtracted by eyeball adjustment of amplitude and phase, which was found to be quicker than an approach based on Fourier transforms. The mean value was then subtracted to give a convincing spread of observations similar to a Gaussian distribution. This was then perturbed by additions and subtractions of temperatures of ±0.2 K to 1.2 K in steps of 0.2 K together with a further four 'changes' of zero amplitude. The choice of a random change was made every 21 days. This was done with nine separate independently random sequences. The resulting perturbed temperature signals, which looked very like normal records, were passed to an independent analysis programme which had information only about each of the sequences on which to base a detection of each of the amplitudes. The scatter of the correlation results, mean errors and the standard deviation of the scatter are shown in Figure 12.

The next experiment of the technique was with a real climate model, the HadGAM2, carried by Parkes.[30] He divided the world's oceans into 89 regions of similar area, and multiplied or divided the initial settings for the concentration of cloud condensation nuclei in each region according to separate independently random sequences. Figure 13 shows the scatter of his predictions for changes in precipitation made by each of the different spray sources around the world at a middle-eastern observation station in the Middle East. Clearly precipitation can be varied both up and down and the scatter is acceptably small.

The results show a drying of the Amazon due to spray off Namibia, in line with those predicted by the Hadley Centre work,[27] but also showed the opposite effect from many other regions with a strong effect from the Aleutian islands in the opposite hemisphere.

The full set of the Parkes results is available online.[30] Confirmation of the data with other climate models would be highly desirable but have not yet been published.

10   Costs

Politicians and investors want firm cost estimates at an early stage of the project, even though their opinions about the cost of NOT preventing

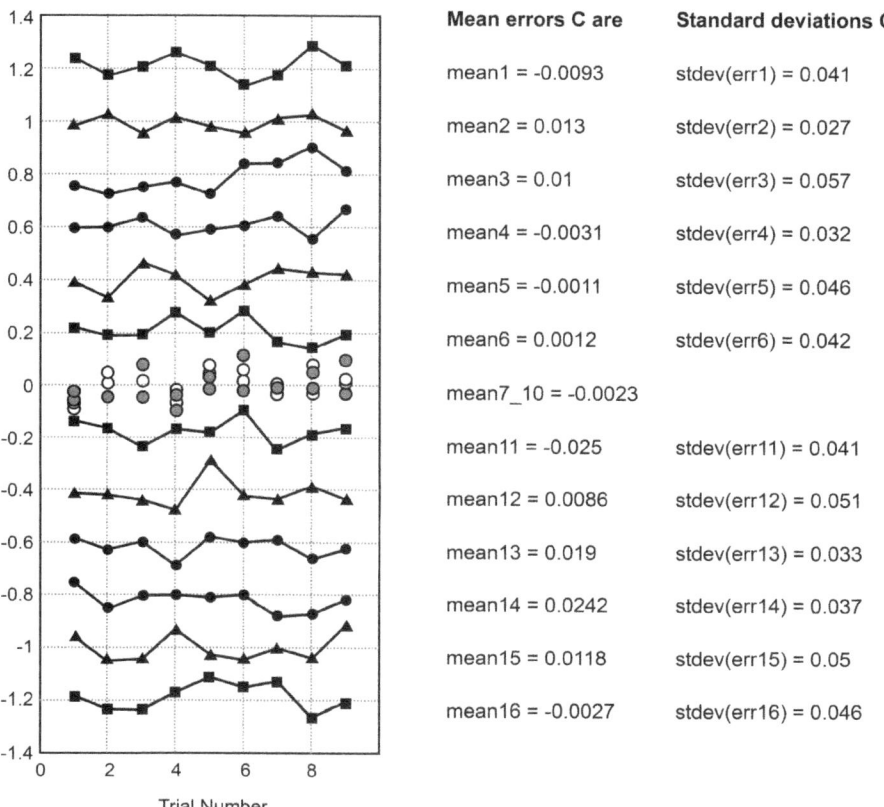

Mean errors C are	Standard deviations C
mean1 = -0.0093	stdev(err1) = 0.041
mean2 = 0.013	stdev(err2) = 0.027
mean3 = 0.01	stdev(err3) = 0.057
mean4 = -0.0031	stdev(err4) = 0.032
mean5 = -0.0011	stdev(err5) = 0.046
mean6 = 0.0012	stdev(err6) = 0.042
mean7_10 = -0.0023	
mean11 = -0.025	stdev(err11) = 0.041
mean12 = 0.0086	stdev(err12) = 0.051
mean13 = 0.019	stdev(err13) = 0.033
mean14 = 0.0242	stdev(err14) = 0.037
mean15 = 0.0118	stdev(err15) = 0.05
mean16 = -0.0027	stdev(err16) = 0.046

Figure 12 A very expensive thermometer would be needed to detect the difference between the original perturbation and the detected result.

climate change are uncertain. The site at gives a useful list of official estimates.[31] Even when a product is being made and sold, its true cost is likely to be a commercial secret. However, it may be possible to use parametric costing using rules based on the nearest similar products made in similar quantities and adjusted by weight, power rating and inflation.

Most ships are made in quite small numbers. Large private yachts are usually status symbols of wealth. One possible 'nearest similar product' is the Flower class corvette built hurriedly for the Royal Navy in World War II. It was based on the design of a whale catcher by Smiths Dock. In 1940 the cost was £60 000, very little of which went for the comfort of the ship's company.[32]

Corrections for the changed value of money are available from several websites.[33] The mean of four estimates for a corvette built today with UK inflation is £2 613 151 with the highest being £2 844 356.

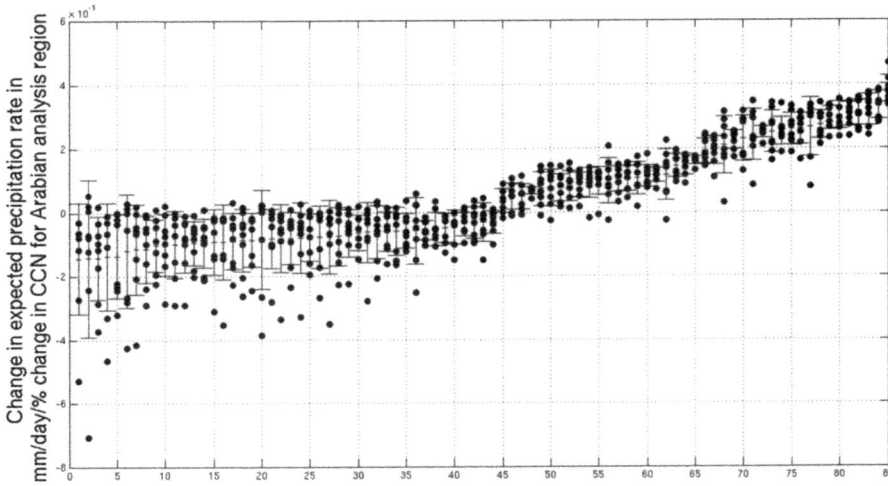

Figure 13   The results from a PhD Thesis by Ben Parkes showing the scatter of precipitation results from eight runs with different coded sequences of the variation of the concentration of condensation nuclei at 89 ocean regions. Precipitation can be changed in both directions.

Table 1   A comparison of corvettes and spray vessels.

	Flower Corvette	Spray vessel	Ratio
Displacement	925 to 1015 tonnes	90 tonnes	10.2
Water line	62.5 m	40 m	1.56
Power	2050 kW	300 kW	6.8
Crew	85	0	
Number built	225	$\sim$300 for 1 W m^{-2}	0.75
Speed	16 knots = 8.23 m s^{-1}	23 knots = 12 m s^{-1}	1.45
Range	3500 miles	unlimited	

The costs of displacement and power will have scaling rules which are a little less than linear. The cost of spray-related components is expected to be about £150 000 per vessel. Even though spray equipment will be novel, writing off the development costs over three spray systems per vessel will give a helpful reduction. Since inventors should never be trusted to publish the costs of their own inventions, readers are invited to pick their own scaling rules, interest rates, write-off fraction and the annual cost of ownership of vessels. Most ships last for about 25 years: the cost estimate of £200 million per year for a cooling of 1 W m^{-2} for marine cloud brightening made by Shepherd *et al.* appears to have written off entire spray vessel fleets rather more frequently.[34] See Table 1 for a comparison of the features of corvettes and spray vessels.

11 Conclusions

Although the engineering problems of marine cloud brightening are formidable there are a number of feasible engineering solutions that have been explored:

- Microfabrication technology can produce billions of orifices per wafer in silicon wafers clad with silicon nitride.
- Piezo excitation of a Rayleigh jet can give a very narrow spread of submicron drop sizes but the capacitance of the piezo-electric elements is large.
- Back-flushing ultrafiltration membranes, used originally to remove 30 nm polio viruses from drinking water and now for pre-filtration in reverse osmosis, can prevent nozzle clogging but nozzles can also be back-flushed.
- The improvement in lift and drag of Flettner rotors hoped for by the addition of Thom fences is not supported by computer modelling.
- It may be possible to use hydrofoil technology with controlled pitch variation of four foils to produce very large reductions in hull drag and several hundred kilowatts of on-board power generation.
- The complexity of parallelogram linkages for moving amas is offset by reductions in stresses in the akas and vaka due to shock loading as in vehicle suspensions.
- Comparisons with inflation corrected cost of WWII corvettes suggest that spray vessels built in similar quantities might cost about £2 million each.
- If the Schwarz and Slingo interpretation of Twomey's work is correct, then a few hundred vessels will be able to correct the thermal effects from pre-industrial times at an annual cost below that of world climate conferences.
- The short lifetime of condensation nuclei allows the use of tactical marine cloud brightening, with times and places for spraying chosen with regard to the phase of monsoons, observations of sea surface temperatures, Arctic ice and jet stream positions. Intelligent control will be much better than steady spray.
- It may be possible to borrow a technique used in signal processing with random sequence variation of the concentration of condensation nuclei, followed by the correlation of each sequence with model results, to get an everywhere-to-everywhere transfer function of spray. This could let us make wet places drier and dry places wetter.

Acknowledgements

Lowell Wood encouraged the work on filtration. Frans Knops of Pentair provided valuable data on ultra-filtration. Work on marine cloud brightening at Edinburgh University is privately supported.

References

1. J. Latham, Control of global warming, *Nature*, 1990, **347**, 339–340.
2. Edinburgh University School of Engineering; http://www.see.ed.ac.uk/~shs/Climate change/Cloud movies/P1000379.MOV, 135 MB video.
3. S. Twomey, Influence of pollution on the short-wave albedo of clouds, *J. Atmos. Sci.*, 1977, **34**, 1149–1152.
4. F. Breon, Aerosol effect on cloud droplet size monitored from satellite, *Science*, 2002, **295**, 834–837.
5. S. E. Schwartz and A. Slingo, Enhanced shortwave radiative forcing due to anthropogenic aerosols, in *Clouds Chemistry and Climate, ed. P. Crutzen and V. Ramanathan*, Springer, Heidelberg, 1996, pp. 191–236.
6. M. H. Smith, P. M. Park and I. E. Consterdine, North Atlantic aerosol remote concentration measured at a Hebridean site, *Atmos. Environ. Part A*, 1991, **25**(3–4), 547–555.
7. R. Bennartz, Global assessment of marine boundary layer cloud droplet number concentration from satellite, *J. Geophys. Res*, 2007, **112**, 1–16.
8. J. E. Kristjansson, Sensitivity to deliberate sea salt seeding of marine clouds-observations and model simulations, *Atmos. Chem. Phys.*, 2012, **12**, 2795–2807.
9. A. Neukermanns, G. Cooper, J. Foster, A. Gadian, L. Galbraith, J. Sudhanu, L. Latham and B. Ormand, Sub-micrometre salt aerosol production: marine cloud brightening, *J. Atmos. Sci.*, 2013, in press.
10. W. van Hoeve, S. Gekle, J. H. Snoeijer and M. Versluis, Breakup of diminutive Rayleigh jets, *Phys. Fluids*, 2010, **22**, 1–11.
11. A. Hashim, A. J. Kordes and S. C. J. M. van Hoof, The effect of ultra-filtration as pretreatment to reverse osmosis in wastewater reuse and seawater desalination applications, *Desalination*, 1999, **124**, 231–242.
12. S. Salter, J. Latham and J. G. Sortino, Sea-going hardware for the cloud albedo method of reversing global warming, *Philos. Trans. R. Soc. London*, 2008, **366**, 3989–4006.
13. J. Seifert, A review of the Magnus effect in aeronautics, *Progr. Aerospace Sci.*, 2012, **55**, 17–45.
14. J. Eggers, Nonlinear dynamics and breakup of free-surface flows, *Rev. Modern Phys.*, 1997, **69**, 865–929.
15. E. Cunningham, On the velocity of steady fall of spherical particles through a fluid medium, *Proc. R. Soc. London, Ser. A*, 1910, **83**, 357–365.
16. A. Thom, *Effects of Disks on the Air Forces on a Rotating Cylinder*, Aeronautical Research Council, R&M, 1934, 1623.
17. J. Norwood, *Performance Prediction for 21st Century Multihull Sailing Yachts*, Amateur Yacht Research Association, London, UK, 1991.
18. S. Mittal and B. Kumar, Flow past a rotating cylinder, *J. Fluid. Mech.*, 2003, **476**, 303–334.
19. T. J. Craft, H. Iacovedes and N. Johnson, and B. E. Launder, Back to the future: Flettner Thom rotors for maritime propulsion?, *Turbulence Heat Mass Transfer*, 2012, 7, 1–10.

20. M. B. Glauert, The flow past a rapidly rotating cylinder, *Proc. R. Soc. London, Ser. A*, 1957, **242**, 108–115.
21. www.skf.com/skf/productcatalogue/jsp/search/advancedSearchDesignationForm2.jsp?lang = en&catalogue = 1.
22. A. Thebault, Hydroptere; www.hydroptere.com/.
23. www.citroenet.org.uk/miscellaneous/hydraulics/hydraulics-1.html.
24. S. Salter, J. T. M. Taylor and N. J. Caldwell, Power conversion mechanisms for wave energy, *Proc. Inst. Mech. Eng, J. Eng. Marine Environ., Part M*, 2002, 1–27.
25. www.artemisip.com/.
26. S. M. Cadwell, R. A. Merrill, C. M. Sloman and F. L Yost, Dynamic fatigue life of rubber, *Ind. Eng. Chem.*, 1940, **12**, 19–23.
27. A. Jones, J. Haywood and O. Boucher, Climate impacts of geoengineering marine stratocumulus clouds, *J. Geophys. Res.*, 2009, **114**, 1–9, doi: 10.1029/2008JD011450.
28. P. J. Rasch, J. Latham and C. C. Chen, Geoengineering by cloud seeding: influence on sea ice and climate system, *Environ. Res. Lett.*, 2009, **4**, 045112–045119.
29. G. Bala, K. Caldeira, R. Nemani, L. Cao, G. Ban-Weiss and H.-J. Shin, Albedo enhancement of marine clouds to counteract global warming: impacts on the hydrological cycle, *Climate Dynamics*, 2011, **37**, 915–931.
30. B. Parkes, *Climate Impacts of Marine Cloud Brightening*, PhD thesis, University of Leeds, 2012; http://homepages.see.leeds.ac.uk/~lecag/parkes.dir/ or http://www.see.ed.ac.uk/~shs/Climate change/Ben_Parkes_Thesis_Final.pdf (accessed 13th January, 2014).
31. www.global-warming-forecasts.com/cost-climate-change-costs-global-warming.php.
32. www.gwpda.org/naval/wcosts.htm.
33. www.bankofengland.co.uk/education/Pages/inflation/calculator/flash/default.aspx.
34. J. Shepherd, Geoengineering the Climate: Science, Governance and Uncertainty, The Royal Society, London, September 2009, p. 35, Table 3.6; http://royalsociety.org/uploadedFiles/Royal_Society_Content/policy/publications/2009/8693.pdf (accessed 13th January 2014).

Stratospheric Aerosol Geoengineering

ALAN ROBOCK

ABSTRACT

In response to global warming, one suggested geoengineering response involves creating a cloud of particles in the stratosphere to reflect some sunlight and cool Earth. While volcanic eruptions show that stratospheric aerosols cool the planet, the volcano analog also warns against geoengineering because of responses such as ozone depletion, regional hydrologic responses, whitening of the skies, reduction of solar power, and impacts of diffuse radiation. No technology to conduct geoengineering now exists, but using airplanes or tethered balloons to put sulfur gases into the stratosphere may be feasible. Nevertheless, it may be very difficult to create stratospheric sulfate particles with a desirable size distribution.

The Geoengineering Model Intercomparison Project, conducting climate model experiments with standard stratospheric aerosol injection scenarios, has found that insolation reduction could keep the global average temperature constant, but global average precipitation would reduce, particularly in summer monsoon regions around the world. Temperature changes would also not be uniform; the tropics would cool, but high latitudes would warm, with continuing, but reduced sea ice and ice sheet melting. Temperature extremes would still increase, but not as much as without geoengineering. If geoengineering were halted all at once, there would be rapid temperature and precipitation increases at 5–10 times the rates from gradual global warming. The prospect of geoengineering working may reduce the current drive toward reducing greenhouse gas emissions, and there are concerns about commercial or military control. Because geoengineering cannot

Issues in Environmental Science and Technology, 38
Geoengineering of the Climate System
Edited by R.E. Hester and R.M. Harrison
© The Royal Society of Chemistry 2014
Published by the Royal Society of Chemistry, www.rsc.org

safely address climate change, global efforts to reduce greenhouse gas emissions and to adapt are crucial to address anthropogenic global warming.

1 Introduction

On September 27, 2013, the Intergovernmental Panel on Climate Change (IPCC) Working Group I released the *Summary for Policymakers of the Fifth Assessment Report*, which stated that "It is *extremely likely* that human influence has been the dominant cause of the observed warming since the mid-20th century." "Extremely likely" is defined as with a greater than 95% probability of occurrence, using the expert judgment of the IPCC scientists. Furthermore, they outlined the projected global warming, sea level rise, changes in precipitation patterns, increase in tropical storms, and other responses to future anthropogenic pollution with a greater degree of certainty than before.

The United Nations Framework Convention on Climate Change (UNFCCC) was established in 1992. Signed by 194 countries and ratified by 189, including the United States, it came into force in 1994. It says in part, "The ultimate objective of this Convention … is to achieve … stabilization of greenhouse gas concentrations in the atmosphere at a level that would prevent dangerous anthropogenic interference with the climate system." "Dangerous anthropogenic interference" was not defined when the UNFCCC was signed, but following the Conference of the Parties in Copenhagen in 2009, the countries of the world agreed that global warming of 2 K above pre-industrial levels should be considered dangerous.

In light of the failure of society to take any concerted actions to deal with global warming in spite of the UNFCCC agreement, two prominent atmospheric scientists published papers in 2006 suggesting that society consider geoengineering solutions to global warming.[1,2] Although this was not a new idea,[3,4] this suggestion generated much interest in the press and in the scientific community, and there has been an increasing amount of work on the topic since then.

The term "geoengineering" has come to refer to both carbon dioxide removal and solar radiation management (SRM),[5,6] and these two different approaches to climate control have very different scientific, ethical and governance issues. This chapter will only deal with solar radiation management, and will focus on the suggestion of producing stratospheric clouds to reflect sunlight in the same way large volcanic eruptions do. Stratospheric aerosols, sunshades in space (see Chapter 8), and marine cloud brightening (see Chapter 6) are the only schemes that seem to have the potential to produce effective and inexpensive large cooling of the planet,[6] but each of them has serious issues, and no such technology currently exists for any of these proposed schemes. Unless otherwise noted, this chapter will use the term "geoengineering" to refer to SRM with stratospheric aerosols.

Clearly, the solution to the global warming problem is mitigation (reduction of emissions of gases and particles that cause global warming, primarily CO_2). Society will also need to adapt to impacts that are already occurring. Whether geoengineering should ever be used will require an analysis of its benefits and risks, as compared to the risks of not implementing it. While research so far has pointed out both benefits and risks from geoengineering, and that it is not a solution to the global warming problem, at some time in the future, despite mitigation and adaptation measures, society may be tempted to try to control the climate to avoid dangerous impacts. Much more research on geoengineering is needed so that society will be able to make informed decisions about the fate of Earth, the only planet in the universe known to sustain life.

This chapter will first discuss how it might be possible to create a permanent cloud in the stratosphere. Next it will survey climate model simulations that inform us of some of the benefits and risks of stratospheric geoengineering. Since full implementation of geoengineering to test these theoretical calculations might be dangerous, lessons from volcanic eruptions, the closest natural analog to stratospheric geoengineering, are used to inform the model results. The next section discusses the ethical and governance aspects of both geoengineering research and potential geoengineering implementation. Finally, the potential benefits and risks of stratospheric geoengineering are summarized.

2   How to Create a Stratospheric Cloud

2.1   Why the Stratosphere?

Every so often, large volcanic eruptions inject massive amounts of sulfur dioxide (SO_2) gas into the stratosphere, the layer of the atmosphere from about 12 km up to 50 km, which resides above the troposphere where we live. The SO_2 is oxidized in the atmosphere to sulfuric acid which has a low enough vapor pressure to form a cloud of droplets. Only volcanic eruptions that are strong enough to get sulfur into the stratosphere have an important impact on climate. They do this by scattering some of the incoming sunlight back to space, thus cooling the surface.[7]

A stratospheric volcanic cloud lasts for a couple years if the eruption is in the Tropics, but for several months if the eruption is at high latitudes. The stratosphere has little vertical motion and no precipitation, so the main removal mechanism is gravitational settling until the particles fall into the troposphere. Initial growth of the particles by coagulation depends on their concentration, and the larger particles fall faster and are removed more rapidly. At the same time, stratospheric circulation moves the particles poleward. The main location for the removal of sulfate from the stratosphere to the troposphere is in the jetstream region in the middle latitudes.[8] The troposphere has vertical motion, mixing, and rain, which can wash particles out of the atmosphere in about a week. The removal of particles from the stratosphere typically is an exponential process. The e-folding time is about one year, which means that a year after the formation of volcanic sulfate particles from tropical injection, the concentration is about 1/3 of the original amount, and after another year, the concentration is about 1/3 of that. For geoengineering, injection would have to be repeated frequently to maintain a stratospheric cloud.

The main suggestion of how to create a stratospheric cloud to reflect sunlight has been to emulate volcanic eruptions.[1-6] Materials other than sulfur have been suggested, for example soot, but soot would be terribly damaging to stratospheric ozone because it would absorb sunlight, heating the stratosphere, and enhancing ozone destruction reactions.[9] This would produce large enhancements of dangerous ultraviolet (UV) flux to the surface. Other substances may be developed in the future, such as minerals or engineered particles,[10] but current work has focused on sulfuric acid.

While sulfuric acid in high concentrations can be dangerous, and acid rain in the troposphere is mainly sulfuric and nitric acid, the amount of annual sulfur emissions to the stratosphere that have been proposed, 5–10 Tg ($Tg = 10^{12}$ g), is much less than the annual volcanic SO_2 emissions into the troposphere,[11] about 13 Tg, plus the annual human emission of SO_2 as a byproduct of burning fossil fuels, about 100 Tg. Nevertheless, sulfur emissions at the level proposed for stratospheric geoengineering would still produce additional impacts on human health and ecosystems.

Since the sulfuric acid clouds created in the stratosphere immediately start to fall out, geoengineering would require continuous replenishment of the sulfur. We know from observations and climate model simulations of volcanic eruptions like the Mt. Pinatubo eruption in the Philippines in 1991, the largest of the 20th Century, that sulfuric acid clouds gradually move from the Tropics poleward covering the entire globe. Therefore, to achieve the longest lifetime for an artificial geoengineering cloud, it would be optimal to start it out in the Tropics. The boundary between the troposphere and the stratosphere, called the tropopause, however, has a maximum altitude in the Tropics, about 18 km. So to conduct stratospheric geoengineering, the task would be to inject sulfur about 20 km into the atmosphere every year in the Tropics. The amount would depend on the size of the effect desired (where to set the planetary thermostat), an unresolved issue.

2.2 Means of Stratospheric Injection

How would it be possible to get several Tg of S into the tropical stratosphere every year? If it were lofted as H_2S gas, with a molecular weight of 34 g per mole S, it would take a little more than half the mass of lofting the S as SO_2 gas, with a molecular weight of 64 g per mole S. The H_2S would probably quickly oxidize to SO_2 and then convert to H_2SO_4. One issue is that H_2S is rather nasty stuff, and even SO_2 can be dangerous, but assuming that industrial procedures could be created to get either gas into a delivery system, what would be the cheapest one?

The first quantitative estimates of the cost for stratospheric geoengineering considered naval guns, hydrogen and hot air balloons, and airplanes for delivering aluminum oxide particles, reflective stratospheric balloons, or soot to the stratosphere,[12] but all options considered were quite expensive. More recent analyses showed that either existing military airplanes or specially designed ones, perhaps pilotless, could deliver 1 Tg S to the tropical lower stratosphere for a few billion US dollars per year.[13,14] While some with experience in scientific aviation question these estimates, it seems that cost would not be a limiting factor if the world was determined to do geoengineering. Towers or tethered balloons have also been suggested,[15] and tethered balloons would be cheaper than airplanes. Figure 1 illustrates some of the suggested options.

Figure 1 Proposed methods of stratospheric aerosol injection, including: airplanes, artillery, balloons and a tower. A mountain top location would require less energy for lofting to stratosphere.
(Drawing by Brian West, Figure 1 from ref. 13).

2.3   Creating an Effective Sulfuric Acid Cloud

An ideal particle would be effective at scattering sunlight, would not affect stratospheric chemistry, and would be safe when it fell out of the stratosphere.[10] As volcanic eruptions provide us with natural examples, sulfate particles are the most studied candidates. A one-time stratospheric injection of SO_2 from a volcanic eruption results in sulfate aerosols with an effective radius of about 0.5 μm, which would be very effective at back-scattering a portion of the incoming sunlight, cooling the surface. Climate model simulations of the impacts of geoengineering (see section 3) assume that the aerosol cloud that would be produced would have properties similar to these volcanic clouds, such as observed after the 1991 Pinatubo eruption. However, if SO_2 were continuously injected into the lower stratosphere, theory says that rather than producing more small particles, much of the SO_2 would be incorporated into existing particles, making them larger.[16] The result is that, per unit mass, the S would be much less effective at scattering sunlight and cooling the surface, and to achieve the same optical depth or reduction in incoming sunlight, as much as 10 times or more mass of S would be needed, if it were possible at all.

This self-limiting feature of stratospheric sulfate aerosols has prompted suggestions of injecting sulfuric acid directly rather than SO_2 to prevent the particle growth,[17,18] but only by widely spreading out the injection of either SO_2 or sulfuric acid would this growth be limited.[19] A system to inject S throughout broad latitude bands has not been developed, and it is not clear that even this would work once there was an existing sulfate cloud, so there is doubt about claims that this would be cheap and easy, since the technology to do stratospheric geoengineering does not currently exist.

The size of aerosol particles not only affects their lifetimes and effectiveness at reflecting sunlight, but it also affects their chemical interactions that destroy ozone. Ozone in the stratosphere absorbs UV radiation from the Sun, protecting life at the surface. Anthropogenic chlorine in the stratosphere, a result of chlorofluorocarbon use in the troposphere (which is now severely limited by the Montreal Protocol and subsequent treaties), is typically found as chlorine nitrate and hydrochloric acid. However, when polar stratospheric clouds form every spring over Antarctica, heterogeneous reactions on the surface of cloud droplets liberate chlorine gas from the reaction between chlorine nitrate and hydrochloric acid, and it catalytically destroys ozone, producing the annual Ozone Hole. Ozone depletion by the same mechanism occurs at the North Pole, but because stratospheric winds are more variable, the vortex does not get as cold, and ozone depletion is more episodic and not as large. As the chlorine concentration in the stratosphere gradually declines, the Ozone Hole is expected to stop forming in 2050 or 2060. The presence of an anthropogenic aerosol cloud as the result of geoengineering, however, would allow ozone depletion to go on even without polar stratospheric clouds. Calculations show that the Ozone Hole would persist for two or three decades more in the presence of geoengineering, and would even start forming in the Northern Hemisphere in cold winters.[20] This effect has been observed after large volcanic eruptions.[21]

3   Climate Impacts of Stratospheric Geoengineering

Although we can learn much from observations of the climatic response to large volcanic eruptions, they are rare and an imperfect analog: volcanic eruptions inject a large amount of SO_2 once; ash is sometimes associated with the sulfate; volcanic eruptions are rare and we have imperfect observations of past ones; and the injection is into a pristine stratosphere and not one with an existing cloud. Therefore some of the processes associated with the continuous creation of a sulfate cloud cannot be studied by observations of volcanic eruptions. The preferred tool for investigating the effects of geoengineering on climate is the climate model. If a climate model has been evaluated by simulations of past volcanic eruptions for which we do have observations and simulations of other causes of climate change, we gain confidence in its ability to simulate similar situations.

3.1   Climate Models

General circulation models (GCMs) of the atmosphere and ocean are the workhorse of the climate community for studying how the climate responds to a large number of natural and anthropogenic forcings (factors that change the amount of energy being received by the climate system). A typical GCM divides the atmosphere and ocean each up into a number of grid boxes and layers, with a typical horizontal spacing of 100 km in the atmosphere and 50 km in the ocean, with 25–90 layers in the atmosphere and 30–40 layers in the ocean. A GCM is started with a particular state of the atmosphere and ocean, and then moves forward in time calculating all the variables of the climate, including wind, ocean current, temperature, clouds, precipitation, sea ice, and amount of sunlight. Modern GCMs also include models of vegetation and the carbon cycle, with interactions on Earth's surface with soil moisture and plants.

GCMs are the same as computer models that are used every day to forecast the weather. However, because they are run for long periods of time, they also explicitly calculate changes in slow-varying components of the climate system, such as ocean currents and heat content, soil moisture, and sea ice, which are typically kept fixed for weather forecasts. Since the atmosphere is a chaotic system, preventing skillful weather forecasts beyond about two weeks, GCMs simulate possible weather states, but not the evolution of weather that did happen in the past or will happen in the future. For that reason, it is typical to use ensembles of GCM simulations, each started with a different arbitrary state of the weather, and to then calculate statistics of the ensemble to study how the climate will change. However, because the real world only evolves along one particular path, climate models are not expected to simulate the exact future state of the climate, only probability distributions and envelopes of climate states that the real world will be expected to inhabit.

3.2 Scenarios of Geoengineering

As with studies of global warming, specific scenarios of geoengineering implementation are needed to conduct studies of the climate impacts. Stratospheric geoengineering has been implemented in GCM studies mainly in two different ways. One is to simply reduce insolation (the solar radiation that reaches the Earth's surface), which is easily implemented in a climate model by reducing the solar constant, or reducing insolation in certain regions. Another scenario is to more realistically simulate the emission of SO_2 gas in the lower stratosphere, and allow models that include these processes to convert the SO_2 to sulfate aerosols, transport the aerosols through the climate system, interacting with sunlight and heat radiation from the Earth along the way, and then remove the aerosols from the system. When aerosols interact with radiation, they alter atmospheric circulation, which then can affect the lifetime and deposition fate of the sulfur.

The specific global warming scenario that stratospheric geoengineering is attempting to address will have a big impact on the resulting climate response. The specific goal of geoengineering will also affect the response. This touches on the larger scale question of, "Whose hand will be on the planetary thermostat?" That is, what is the goal of geoengineering? Is it to keep the global average temperature constant at the value at the time of geoengineering implementation? Is it to only allow warming up to the predetermined level of dangerous anthropogenic interference, say 2 K above pre-industrial temperatures? Is it to just slow global warming and compensate for only part of future warming? Or is it to cool the planet back to a level colder than current conditions, since the planet is already too warm, and sea ice melting, sea level rise, and the potential for Arctic methane releases are already dangerous at the current climate?

The impacts of geoengineering also depend on how GCM results are evaluated. Once the goal of geoengineering is decided, how are the resulting climate changes to be judged? As compared to the climate at the time of implementation? As compared to the climate that would have resulted at some time in the future if no geoengineering had been used? As compared to pre-industrial climate?

Early geoengineering GCM experiments each made different choices for each of these factors, and therefore it was not possible to compare the results to see if they were robust with respect to each other, as each was doing different experiments. For example, some tried to just cool the Arctic, and some the entire planet. Some tried to balance a doubling of CO_2 and others compensate for gradually increasing greenhouse gases. To address this issue, the Geoengineering Model Intercomparison Project (GeoMIP) was implemented.[22] GeoMIP developed four scenarios of stratospheric geoengineering, and asked all the GCM modeling groups in the world to conduct the same experiments and share their results so that others could analyze them and compare the effectiveness and risks of geoengineering with respect to a number of different metrics.

The GeoMIP scenarios are shown in Table 1 and Figure 2. These built on experiments already conducted by modeling groups to examine the climate system response to increases of CO_2.[23] G1 and G2 were the easiest to implement, involving adjusting the amount of incoming sunlight to balance the heating caused by an instantaneous quadrupling of CO_2 or a gradual increase of CO_2 of 1% year^{-1}. Twelve modeling groups from around the world participated in the first round of experiments. G3 and G4 were more "realistic," involving a "business-as-usual" scenario of increasing greenhouse gases by modeling the injection of SO_2 into the tropical lower stratosphere to create a global sulfate cloud to either balance the anthropogenic heating or to immediately overwhelm that heating (say in the event of a planetary emergency) and injecting 5 Tg of SO_2 per year. G1 and G2 start from an artificial equilibrium climate, while G3 and G4 start from a more realistic warming climate. This means that for G3 and G4, preventing further radiative forcing would not be enough to stop the planet from warming, since there would be a built-in energy imbalance at the start.

Table 1   A summary of the four GeoMIP experiments. The different experimental designs are shown in Figure 2. RCP4.5 (representative concentration pathway resulting in 4.5 W m^{-2} radiative forcing) is a "business-as-usual" scenario used to force climate models in recent standardized experiments.[23] (Table 1 from ref. 22).

G1	Instantaneously quadruple the CO_2 concentration (as measured from pre-industrial levels) while simultaneously reducing the solar constant to counteract this forcing.
G2	In combination with a 1% increase in CO_2 concentration per year, gradually reduce the solar constant to balance the changing radiative forcing.
G3	In combination with RCP4.5 forcing, starting in 2020, gradual ramp-up the amount of SO_2 or sulfate aerosol injected, with the purpose of keeping global average temperature nearly constant. Injection will be done at one point on the Equator or uniformly globally. The actual amount of injection per year will need to be fine tuned to each model.
G4	In combination with RCP4.5 forcing, starting in 2020, daily injections of a constant amount of SO_2 at a rate of 5 Tg SO_2 year^{-1} at one point on the Equator through the lower stratosphere (approximately 16–25 km in altitude) or the particular model's equivalent. These injections would continue at the same rate through the lifetime of the simulation.

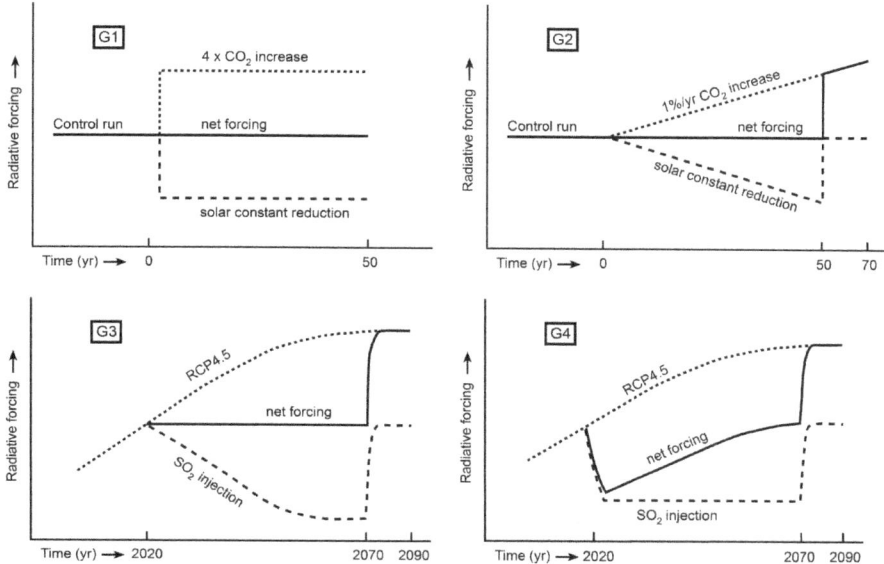

Figure 2   The four GeoMIP experiments, described in Table 1. G1: The experiment is started from a control run. The instantaneous quadrupling of CO_2 concentration from pre-industrial levels is balanced by a reduction in the solar constant until year 50. G2: The experiment is started from a control run. The positive radiative forcing of an increase in CO_2 concentration of 1% year^{-1} is balanced by a decrease in the solar constant until year 50. G3: The experiment approximately balances the positive radiative forcing from the RCP4.5 scenario by an injection of SO_2 or sulfate aerosols into the tropical lower stratosphere. G4: This experiment is based on the RCP4.5 scenario, where immediate negative radiative forcing is produced by an injection of SO_2 into the tropical lower stratosphere at a rate of 5 Tg year^{-1}. (Figures 1–4 from ref. 22).

3.3   Global and Regional Temperature Impacts

While a wide range of potential geoengineering implementations might be considered, the GeoMIP experiments allow the best opportunity to systematically study the climate system response. Since in general the climate system responds linearly to changes in the amount of energy being added or taken away, other scenarios of geoengineering can be scaled by the GeoMIP results for a first order understanding of the climate system response.

Figure 3 shows the global response in 11 different climate models for the G2 experiment.[24] A 1% year^{-1} CO_2 increase (approximately what we have observed in the past several decades) would produce a global warming of about 1 K in 50 years. With varying levels of success, climate models are able to completely stop this warming by reducing sunlight. However, when geoengineering is halted at year 50, the result is rapid global warming, at a rate as much as 10 times the rate we will experience with no geoengineering. It is often the rate of change of climate that is more disruptive than the

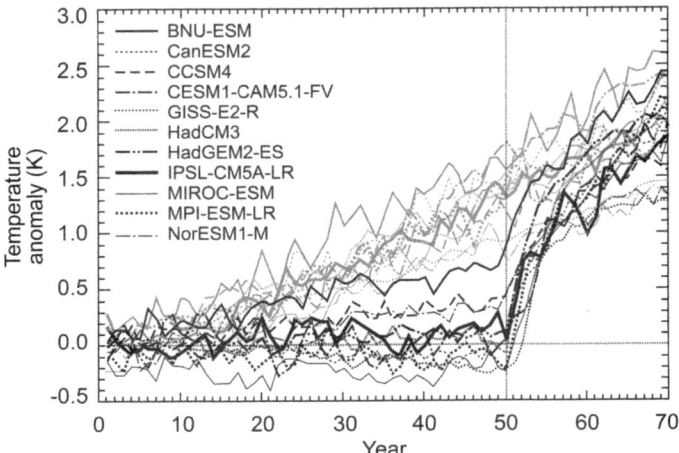

Figure 3 Evolution of annual mean anomaly of global mean near-surface air temperature (K) in the G2 simulations (black lines) with respect to the long-term mean from each model's control simulation. Time series from corresponding 1% CO_2 year^{-1} increase simulations are also shown (gray lines). The termination of geoengineering in the G2 simulations is indicated by the dashed vertical line.
(Figure 1 from ref. 24; see this reference for climate model abbreviations and details).

actual climate, as it is difficult in some cases to quickly adapt, say for infrastructure built under the assumption of no, or gradual, change. And if geoengineering were ever actually implemented, there would be no way to predict when society might lose the will or means to continue the geoengineering, producing this termination effect. While it would be logical to slowly ramp down geoengineering if there were a reason to stop it, it is easy to imagine a devastating drought or flood somewhere in the world that is blamed on geoengineers, with a demand that geoengineering be halted at once.

Even if it were possible to control the global temperature with a global reduction of sunlight, say from tropical sulfur injections, the G1 experiment teaches us that the temperature changes would not be uniform.[25] Figure 4 shows that if the warming from CO_2 were balanced by insolation reduction, keeping the global average temperature from changing, temperatures would fall in the Tropics and continue to go up in the Arctic. The regional details are not well known, however, as indicated by the stippling in the figure. The simple explanation for the variation with latitude is that while the warming from CO_2 is a little bit larger in the Tropics than the poles (because the downward heat radiation from the excess CO_2 is a function of temperature and it is warmer in the Tropics), the warming is still fairly well distributed around the world. However, there is much more sunlight to reflect in the Tropics than at the poles, and the change in energy by blocking sunlight is

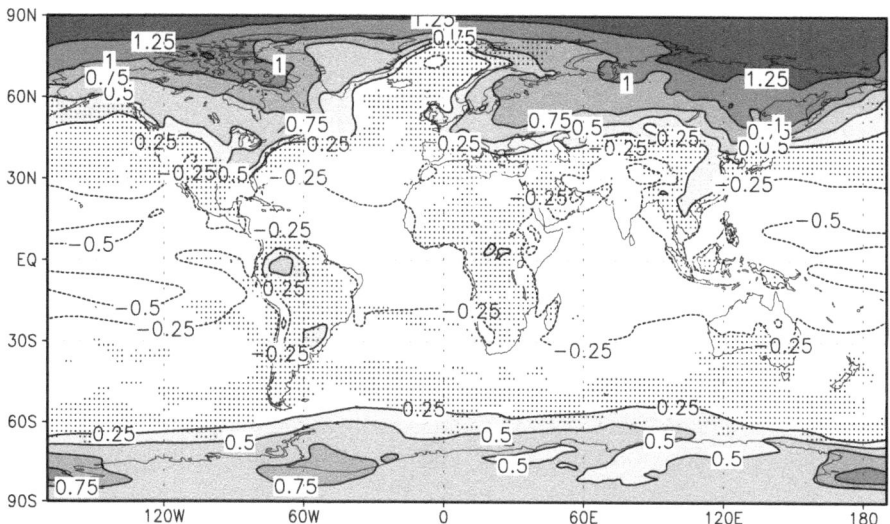

Figure 4 All-model ensemble annual average surface air temperature differences (K) for G1 minus the control run, averaged over years 11–50 of the simulation. Stippling indicates where fewer than 75% of the models (9 out of 12) agree on the sign of the difference. (Figure 2 from ref. 25).

much more asymmetric. This means that if global geoengineering were to be used to try to stop sea level rise, there would have to be global cooling to not only keep the ice sheets at the poles (Greenland and Antarctica) from melting, but also to reverse the built-in sea level rise already happening from energy in the oceans from the warming that has already taken place in the recent past.[26]

3.4 Global and Regional Precipitation and Monsoon Impacts

Temperature is important, as warming directly affects sea level through melting land-based glaciers and ice sheets and expanding the ocean water; reduced seasonal snowpack threatens water supplies; and crops are sensitive to temperature changes. Precipitation changes from global warming are a more direct threat, however, to agriculture and water supplies. One of the aims of geoengineering might be to reverse changes in precipitation patterns being caused by global warming, particularly the expansion of areas of drought. However, volcanic eruptions are known to increase drought in certain monsoon regions.[27] In addition, global warming is producing more precipitation extremes, with the strongest thunderstorms and hurricanes getting stronger, producing more flooding. It turns out that temperature and precipitation changes cannot be controlled independently.

Figure 5 shows global average precipitation changes from the G2 experiment. At the same time that global average temperature is being kept

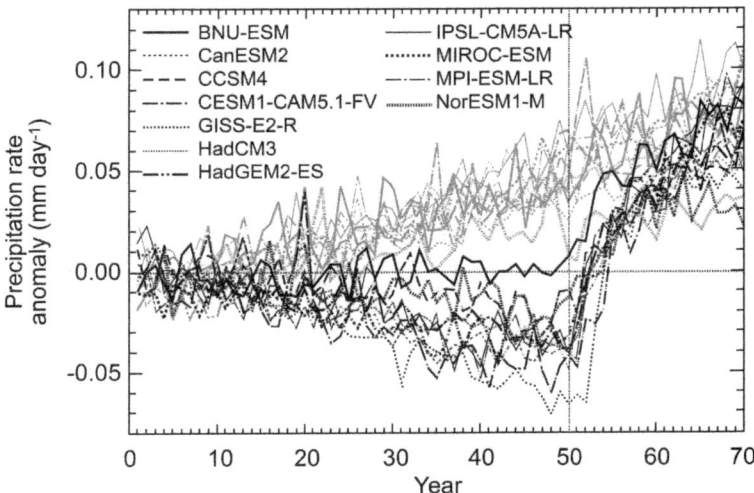

Figure 5   As in Figure 3 but for the anomaly in global mean precipitation rate (mm
day^{-1}).
(Figure 2 from ref. 24; see this reference for climate model abbreviations
and details).

constant by balancing increased CO_2 by insolation reduction (see Figure 3),
global average precipitation would decrease. This result reproduces previous
results and is well-understood.[28] Increases of greenhouse gases, particularly
CO_2, absorb longwave heat radiation throughout the troposphere, decreasing the lapse rate of temperature and making the atmosphere more
stable, reducing precipitation. At the same time they warm the surface,
producing more evapotranspiration and making the hydrological cycle
stronger, increasing precipitation. The evapotranspiration effect wins out
over time, but there is a delay in the increase in precipitation in response to
increases in CO_2, and this can be seen by comparing the gray lines in Figures
3 and 5. While the temperature effect is seen immediately, it takes 10–20
years for the precipitation increases to emerge from the initial values. Insolation reduction only affects the evaporation rate changes from CO_2, but
does not affect the lapse rate part, so it only partially compensates for precipitation changes in a combined high CO_2, low sunlight environment, and
global precipitation therefore goes down.

As impacts are felt locally, the spatial pattern of precipitation changes is
important. The monsoon regions of the world (see Figure 6) are regions
where the difference between summer average and winter average precipitation exceeds 180 mm and the local summer monsoon precipitation produces at least 35% of the total annual rainfall.[29] They are important for
agriculture, particularly in Asia and Africa. In the G1 experiment,[30] summer
land precipitation went up in six of the seven monsoon regions because of
CO_2 increases in the base case, but in six of the seven regions, G1 caused a
reduction of summer land precipitation. (see Figure 7).

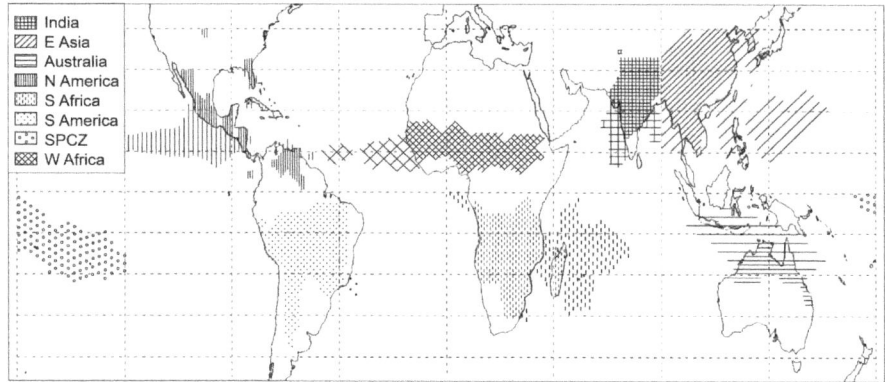

Figure 6 Monsoonal regions (shaded) over land (more dense shading) and ocean (less dense shading), derived from the *Global Precipitation Climatology Project (GPCP) dataset,*[58] covering the years 1979 to 2010, and using criteria described in ref. 29. The North and South American monsoon are defined here as the American monsoon north and south of the equator, respectively.
(Figure 6 from ref. 30).

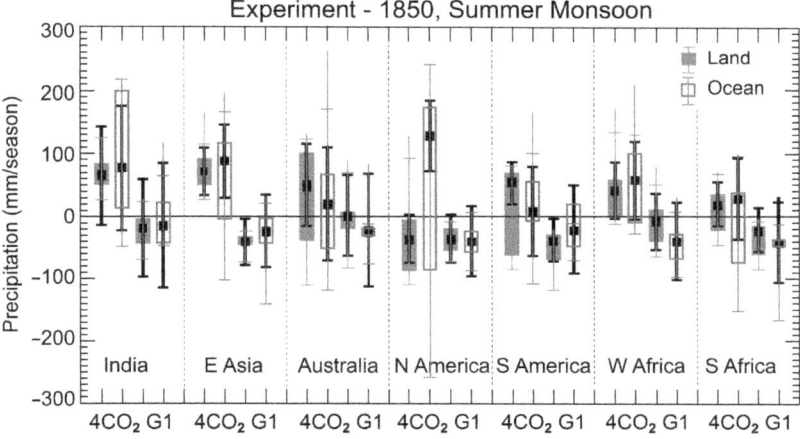

Figure 7 Summer monsoon change of precipitation for $4 \times CO_2$ and G1 with regard to 1850 (control) conditions. Results are for land (grey – 1st and 3rd column for each region) and ocean (white – 2nd and 4th column for each region) and for different regions (see Figure 6). The multi-model range is illustrated by a vertical line, the 25th and 75th percentile of multi-model results are given as a box, and the 5th and 95th percentile are horizontal bars. In addition, the multi-model median is shown as solid symbols and the inter-annual variability of each experiment, represented by the median standard deviation of seasonal averages for each model, is show as error bars pointing off the median of the multi-model results. The two left whisker plots for each region are the $4 \times CO_2$ statistics, and the two rightmost whiskers plots are for G1.
(Figure 14 from ref. 30).

Whether this reduction of summer monsoons would have a large impact on agriculture would depend on how evapotranspiration changed, how much CO_2 fertilization (increased photosynthesis and plant growth as CO_2 concentration rises) would compensate for the negative impacts of geoengineering,[31,32] and how humans would adapt to the changing climate. In G1, evapotranspiration reductions partially compensated for precipitation reductions over most of the land areas.[25,30] Net primary productivity (a measure of natural and managed biological productivity) changes from geoengineering are not well known, as there is a large variation in model responses depending on how the models considered the effects of CO_2 fertilization.[24,25] Much more work is needed on the biological response to stratospheric geoengineering, including modeling the effects on specific species from the range of changes that would result, before we can have a definitive answer.

3.5   Impacts of Enhanced Diffuse Radiation

Among the many potential risks associated with stratospheric geoengineering,[33] is the impact of more diffuse and less direct radiation on the surface of Earth. Much of the light impinging on a stratospheric aerosol cloud would be forward scattered, producing enhanced diffuse radiation, which means that the sky will appear whiter due to the perpetual thin cloud there.[34] In addition to no more blue skies, with its as yet unquantified psychological impact on everyone on Earth, this redistribution of direct radiation to diffuse would have impacts on solar generation of electricity and on the biosphere.

While photovoltaic solar panels are currently the most ubiquitous way that electricity is generated with sunlight, those that focus the direct solar beam with mirrors and boil water or other fluids to drive turbines are more efficient at using solar power. After large volcanic eruptions, observations at Mauna Loa, Hawaii, have shown a large decrease in this direct radiation, for example by 34% after the 1982 El Chichón eruption, which put about 7 Tg of SO_2 into the stratosphere.[7] After the 1991 Mt. Pinatubo eruption, during the summer of 1992 in California when the effects of the eruption were the strongest, solar generators using direct solar radiation produced 34% less electricity than during the period with a clean stratosphere.[35] While the correspondence of these numbers is fortuitous, they point out that one unintended consequence of geoengineering would be a reduction of electricity generation from one of the key sources needed to mitigate the emission of CO_2.

In general, plants grow more when subject to more diffuse light.[13] Stomata on leaves can stay open longer when the leaves are not as hot, as this reduces the loss of water when they are open to obtain CO_2 for photosynthesis. In addition, diffuse light can penetrate the canopy, also increasing photosynthesis. The result is that the CO_2 sink at the surface would increase with geoengineering. In fact, a reduction of the rate of CO_2 increase has been observed in the Mauna Loa CO_2 record for about a year after each of the large volcanic eruptions since the record was started: Agung in 1963, El Chichón in 1982, and Pinatubo in 1991. A calculation of net primary productivity after the Pinatubo eruption, accounting for the effects of changes of temperature and precipitation and isolating the diffuse radiation effect, found a 1 Pg C increase in the CO_2 sink in 1992 (ref. 36), more than 10% of the current annual anthropogenic carbon input to the atmosphere. While an increased carbon sink would be a benefit of stratospheric geoengineering, the effect would be felt differentially between different plant species, and whether it would help or hurt the natural ecosystem, or whether it would preferentially favor weeds rather than agricultural crops, has not been studied in detail yet.

4 Ethics and Governance of Stratospheric Geoengineering

The audacious idea of actually controlling Earth's climate brings up a number of ethical and governance issues. The fundamental question is that of where to set the planet's thermostat. Who would decide how to carry out geoengineering? What values would be used to decide? For whose benefit would this decision be made? For those controlling the geoengineering? For the entire planet, however defined? For the benefit of those most at risk? For only humans, or taking into account the rest of the natural biosphere? These decisions are in the realms of politics and power, and are different from testable scientific hypotheses, but scientific evaluations of the benefits, risks, and uncertainties of various proposals should, in an ideal world, inform decisions about implementation of geoengineering. The discussion in this section separates the issues of research and deployment, and speculates about international governance.

4.1 Ethics and Governance of Research

There have been many recent recommendations that geoengineering research be enhanced, including from the UK Royal Society,[5] the American Meteorological Society,[37] the American Geophysical Union,[38] the U. S. Government Accountability Office,[39] and prominent scientists.[40,41] But is such research ethical?[42] Does it lead to a slippery slope toward geoengineering deployment? Does it take resources away from other more useful pursuits? Is it yet another way for developed countries to continue to dominate the world to benefit themselves? Does the knowledge that this research is ongoing present a "moral hazard,"[43] and reduce whatever political drive there is toward mitigation, since it will be seen as an easier solution to global warming? Does indoor geoengineering research (in a laboratory or a computer, with no emissions to the environment) have different ethical issues from outdoor research (in which sulfur is emitted into the stratosphere to test potential technology and its impacts)? Are weapons being developed in the guise of understanding the science of geoengineering, which was a strong motivation for past research on weather and climate modification?[44] Or would it be unethical not to investigate a technology that may prevent widespread dangerous impacts on climate associated with global warming? Would it be unethical not to be able to provide policymakers in the near future with detailed information about the benefits and risks of various geoengineering proposals so that they can make informed decisions about implementation? Would it be unethical not to develop the technology to carry out geoengineering, both so that the costs and efficacy can be determined (maybe it will prove impossible or much too expensive or dangerous), and to have the designs available so that it could be rapidly implemented if needed?

Answers to these questions are summarized here, based on a longer article.[42] Additional concerns about geoengineering research include the fact that the existence of the technology might enable hasty, politically-driven decisions to be deployed. And as a recent report says,[45] "SRM research could constitute a cheap fix to a problem created by developed countries, while further transferring environmental risk to the poorest countries and the most vulnerable people." The same report also discusses hubris, "Artificial interference in the climate system may be seen as hubristic: 'playing God' or 'messing with nature,' which is considered to be ethically and morally unacceptable. While some argue that human beings have been interfering with the global climate on a large scale for centuries, SRM involves *deliberate* interference with natural systems on a planetary scale, rather than an inadvertent side effect. This could be an important ethical distinction."[45]

If the research itself were dangerous, directly harming the environment, this would bring up ethical concerns. Is it ethical to create additional pollution just for the purpose of scientific experiments? There have been no such outdoor experiments in the stratosphere. To test whether there were a climate response or whether existing sulfuric acid cloud droplets would grow in response to additional emissions would require very large emissions, essentially implementation of geoengineering,[46] and would therefore be unethical. But what about flights to spray a little SO_2 or other S species and then observe how particles grow or the response of ozone? Although no such governance now exists, any such outdoor experiments need to be evaluated by an organization, like a United Nations commission, independent from the researchers, that evaluates an environmental impact statement from the researchers and determines that the environmental impact would be negligible, as is done now for emissions from the surface. There would also need to be enforcement of the limits of the original experiment, so that it would not be possible to emit a little more, or over a larger area or for a longer time than in the initial plans, should the experimenters be tempted to expand the experiment in light of inconclusive results.

To make decisions about ethics requires a declaration of values, unlike in the physical sciences, where nature follows well-accepted laws, such as conservation of energy. The above conclusions are based on the following principles: (1) curiosity-driven indoor research cannot and should not be regulated, if it is not dangerous; (2) emissions to the atmosphere, even for scientific purposes, should be prohibited if they are dangerous; and (3) the idea of geoengineering is not a secret, and whatever results from it will need to be governed the same way as all other dangerous human inventions, such as ozone depleting substances and nuclear weapons.

The conclusions are therefore, "in light of continuing global warming and dangerous impacts on humanity, indoor geoengineering research is ethical and is needed to provide information to policymakers and society so that we can make informed decisions in the future to deal with climate change. This research needs to be not just on the technical aspects, such as climate change and impacts on agriculture and water resources, but also on historical precedents, governance, and equity issues. Outdoor geoengineering research, however, is not ethical unless subject to governance that protects society from potential environmental dangers...Perhaps, in the future the benefits of geoengineering will outweigh the risks, considering the risks of doing nothing. Only with geoengineering research will we be able to make those judgments."[42]

4.2   Ethics and Governance of Deployment

Suppose that technology is developed to produce an effective stratospheric aerosol cloud using sulfur or more exotic materials, and that estimated annual direct costs are in the order of US $10 000 000 000. Considering that this is less than $\frac{1}{4}$ of the annual *profits* of one of the leading purveyors of products that emit greenhouse gases, ExxonMobil, it would be very tempting to implement – global warming problem solved! But what about the risks? Would the prevention of more severe weather, crop losses, and sea level rise be worth the negative impacts geoengineering would have in some regions? Would it be OK to allow continued ocean acidification, and its impact on ocean life? Could we be sure that there would be no sudden termination of geoengineering, with its associated rapid climate change?

How would the world make this decision?[47] How would it be possible to determine that we have reached a point where there is a planetary emergency? By what criteria, and an emergency for whom? Even if we could have an accurate idea of the losers of such a decision, how well would society compensate them for the disruption to their livelihoods and communities? The past record of such relief is not good – just think of what happens when "development" destroys old neighborhoods or people are moved when a dam is built. And given the natural variability of weather and climate, how would it even be possible to attribute negative events to the geoengineering? What if a country or region had either severe flooding or severe drought for a couple years in a row during the summer monsoon? Although it would not be possible to definitively point the finger at geoengineering, certainly such claims would be made, and there would be demands not only for compensation, but also for a halt to geoengineering.

In medical procedures, the principle of "informed consent" applies. How could society get informed consent from the entire planet? Would all governments of the world have to agree? What if they agree to control the climate, but some want the temperature to be a certain value and others a different one? Would this result in international conflict? Or what if a big multinational geoengineering corporation is running things? They would have an interest in continuing the work no matter what, and would argue that we cannot stop because it will kill jobs. The over-built militaries of the world, particularly in the United States, are a lesson in how dangerous technologies perpetuate themselves. Weapons continue to be built because of lobbying by special interests. Nuclear weapons are the most dangerous example.[48,49]

There have been a number of papers addressing the ethical and governance issues associated with geoengineering,[50–53] and they discuss the above issues and others. One such attempt to do this is the Oxford Principles.[54] They are "geoengineering to be regulated as a public good," "public participation in geoengineering decision-making," "disclosure of geoengineering research and open publication of results," "independent assessment of impacts," and "governance before deployment." While these are only a proposal with no enforcement, there is no evidence that legitimate geoengineering researchers are not attempting to follow them. One of the more interesting papers imagines various scenarios of future developments that result in different decisions about deployment, with different consequences.[55] Given the uncertainty that will remain even after more research is completed, the dangers of human mistakes either in the construction or operation of the technology, and the possibilities of surprises, will society stake the fate of our planet on geoengineering technology?

5   Benefits and Risks of Stratospheric Geoengineering

Stratospheric geoengineering has the potential to reduce some or all of the warming produced by anthropogenic greenhouse gas emissions, which would then lessen or eliminate the dangerous impacts of global warming, including floods, droughts, stronger rainfall events, stronger hurricanes, sea ice melting, land-based ice sheet melting, and sea level rise. But would these benefits reduce more risk from global warming than would be created by the implementation of geoengineering? That is, would implementation of geoengineering lower overall risk to Earth or add to the level of risk? And will research ever be able to answer this question definitively enough for rational policy decisions? Or will some of the less quantifiable risks, such as the threat of conflict due to disagreement on how to control the planet or unknown unknowns, prevent any agreement on governance?[47]

In addition to the risks and benefits discussed above, other risks and benefits have been suggested but have not been quantified.[33,56] These include: the conflict of geoengineering with the *United Nations Convention on the Prohibition of Military or any Other Hostile Use of Environmental Modification Techniques*; the potential of the sulfuric acid to damage airplanes flying in the stratosphere; an increase in sunburn, as people would be less likely to protect themselves from diffuse radiation; the effect of changing UV on tropospheric chemistry; and unexpected benefits that would accompany unexpected consequences. Table 2 summarizes the risks and benefits from stratospheric geoengineering.

Table 2 Benefits and risks of stratospheric geoengineering. The effects that are observed after volcanic eruptions are indicated by an asterisk (*).[56] (Updated from ref. 57).

Benefits	Risks
1. Reduce surface air temperatures*, which could reduce or reverse negative impacts of global warming, including floods, droughts, stronger storms, sea ice melting*, land-based ice sheet melting, and sea level rise* 2. Increase plant productivity* 3. Increase terrestrial CO_2 sink* 4. Beautiful red and yellow sunsets* 5. Unexpected benefits	1. Drought in Africa and Asia* 2. Perturb ecology with more diffuse radiation* 3. Ozone depletion, with more UV at surface* 4. Whiter skies* 5. Less solar energy generation* 6. Degrade passive solar heating 7. Environmental impact of implementation 8. Rapid warming if stopped* 9. Cannot stop effects quickly 10. Human error 11. Unexpected consequences 12. Commercial control 13. Military use of technology 14. Conflicts with current treaties 15. Whose hand on the thermostat? 16. Degrade terrestrial optical astronomy* 17. Affect stargazing* 18. Affect satellite remote sensing* 19. Societal disruption, conflict between countries 20. Effects on airplanes flying in stratosphere* 21. Effects on electrical properties of atmosphere 22. More sunburn (from diffuse radiation) 23. Continued ocean acidification 24. Impacts on tropospheric chemistry 25. Moral hazard – the prospect of it working would reduce drive for mitigation 26. Moral authority – do we have the right to do this?

In the real world, decisions are made without full knowledge, and sometimes under pressure from extraordinary events. In my opinion, much more research in stratospheric geoengineering, transparently and published openly, is needed so that the potential benefits and risks that can be quantified will be known to aid in future policy decisions.

Even at this late date, a global push to rapid decarbonization, by imposing a carbon tax, will stimulate renewable energy, and allow solar, wind, and newly developed energy sources to allow civilization to prosper without using the atmosphere as a sewer for CO_2. Adaptation will reduce some of the negative impacts of global warming. Geoengineering does not now appear to be a panacea, and research in geoengineering should be in addition to strong efforts toward mitigation, and not a substitute. In fact, geoengineering may soon prove to be so unattractive that research results will strengthen the push toward mitigation.

Acknowledgments

I thank Brian West for drawing Figure 1 and Ben Kravitz for drawing Figure 4. Supported by U.S. National Science Foundation grants AGS-1157525 and CBET-1240507.

References

1. P. Crutzen, *Climatic Change*, 2006, **77**, 211.
2. T. M. L. Wigley, *Science*, 2006, **314**, 452.
3. M. I. Budyko, *Climate and Life*, Academic Press, New York, NY, 1974.
4. R. E. Dickinson, *Clim. Change*, 1996, **33**, 279.
5. J. Shepherd *et al.*, *Geoengineering the Climate: Science, Governance and Uncertainty*, Royal Society Policy document 10/09, Royal Society, London, UK, 2009.
6. T. M. Lenton and N. E. Vaughan, *Atmos. Chem. Phys.*, 2009, **9**, 5539.
7. A. Robock, *Rev. Geophys.*, 2000, **38**, 191.
8. B. Kravitz, A. Robock, L. Oman, G. Stenchikov and A. B. Marquardt, *J. Geophys. Res.*, 2009, **114**, D14109, doi: 10.1029/2009JD011918.
9. B. Kravitz, A. Robock, D. T. Shindell and M. A. Miller, *J. Geophys. Res.*, 2012, **117**, D09203, doi: 10.1029/2011JD017341.
10. F. D. Pope, P. Braesicke, R. G. Grainger, M. Kalberer, I. M. Watson, P. J. Davidson and R. A. Cox, *Nature Clim. Change*, 2012, **2**, 713, doi: 10.1038/nclimate1528.
11. H.-F. Graf, J. Feichter and B. Langmann, *J. Geophys. Res.*, 1997, **102**(D9), 10,727, doi: 10.1029/96JD03265.
12. Committee on Science Engineering and Public Policy, Appendix Q, in *Policy Implications of Greenhouse Warming: Mitigation, Adaptation, and the Science Base*, Natl. Acad. Press, Washington, DC, 1992, 433.
13. A. Robock, A. B. Marquardt, B. Kravitz and G. Stenchikov, *Geophys. Res. Lett.*, 2009, **36**, L19703, doi: 10.1029/2009GL039209.
14. J. McClellan, J. Sisco, B. Suarez and G. Keogh, *Geoengineering Cost Analysis*, Aurora Flight Sciences Corp, Cambridge, MA., 2010, AR10-182.
15. P. Davidson, C. Burgoyne, H. Hunt and M. Causier, *Phil. Trans. R. Soc. London, Ser. A*, 2012, **370**, 4263, doi: 10.1098/rsta.2011.0639.
16. P. Heckendorn, D. Weisenstein, S. Fueglistaler, B. P. Luo, E. Rozanov, M. Schraner, L. W. Thomason and T. Peter, *Environ. Res. Lett.*, 2009, **4**, 045108, doi: 10.1088/1748-9326/4/4/045108.

17. J. P. Pinto, R. P. Turco and O. B. Toon, *J. Geophys. Res.*, 1989, **94**(D8), 11,165, doi: 10.1029/JD094iD08p11165.
18. J. R. Pierce, D. K. Weisenstein, P. Heckendorn, T. Peter and D. W. Keith, *Geophys. Res. Lett.*, 2010, **37**, L18805, doi: 10.1029/ 2010GL043975.
19. J. M. English, O. B. Toon and M. J. Mills, *Atmos. Chem. Phys.*, 2012, **12**, 4775, doi: 10.5194/acp-12-4775-2012.
20. S. Tilmes, R. Müller and R. Salawitch, *Science*, 2008, **320**, 1201 doi: 10.1126/science.1153966.
21. S. Solomon, *Rev. Geophys.*, 1999, **37**, 275.
22. B. Kravitz, A. Robock, O. Boucher, H. Schmidt, K. Taylor, G. Stenchikov and M. Schulz, *Atmos. Sci. Lett.*, 2011, **12**, 162, doi: 10.1002/asl.316.
23. K. E. Taylor, R. J. Stouffer and G. A. Meehl, *Bull. Am. Meterol. Soc.*, 2012, **93**, 485, doi: 10.1175/BAMS-D-11-00094.1.
24. A. Jones, J. M. Haywood, K. Alterskjær, O. Boucher, J. N. S. Cole, C. L. Curry, P. J. Irvine, D. Ji, B. Kravitz, J. E. Kristjánsson, J. C. Moore, U. Niemeier, A. Robock, H. Schmidt, B. Singh, S. Tilmes, S. Watanabe and J.-H. Yoon, *J. Geophys. Res. Atmos.*, 2013, **118**, 9743, doi: 10.1002/ jgrd.50762.
25. B. Kravitz, K. Caldeira, O. Boucher, A. Robock, P. J. Rasch, K. Alterskjær, D. Bou Karam, J. N. S. Cole, C. L. Curry, J. M. Haywood, P. J. Irvine, D. Ji, A. Jones, J. E. Kristjánsson, D. J. Lunt, J. Moore, U. Niemeier, H. Schmidt, M. Schulz, B. Singh, S. Tilmes, S. Watanabe, S. Yang and J.-H. Yoon, *J. Geophys. Res. Atmos.*, 2013, **118**, 8320, doi: 10.1002/jgrd.50646.
26. J. Moore, S. Jevrejeva and A. Grinsted, *Proc. Nat. Acad. Sci. U. S. A.*, 2010, **107**(36), 15699, doi: 10.1073/pnas.1008153107.
27. J. M. Haywood, A. Jones, N. Bellouin and D. Stephenson, *Nature Clim. Change*, 2013, **3**, 660, doi: 10.1038/nclimate1857.
28. G. Bala, P. B. Duffy and K. E. Taylor, *Proc. Nat. Acad. Sci. U. S. A.*, 2008, **105**, 7664, doi: 10.1073/pnas.0711648105.
29. B. Wang and Q. Ding, *Geophys. Res. Lett.*, 2006, **33**(6), L06711, doi: 10.1029/2005GL025347.
30. S. Tilmes, J. Fasullo, J.-F. Lamarque, D. R. Marsh, M. Mills, K. Alterskjær, H. Muri, J. E. Kristjánsson, O. Boucher, M. Schulz, J. N. S. Cole, C. L. Curry, A. Jones, J. Haywood, P. J. Irvine, D. Ji, J. C. Moore, D. B. Karam, B. Kravitz, P. J. Rasch, B. Singh, J.-H. Yoon, U. Niemeier, H. Schmidt, A. Robock, S. Yang and S. Watanabe, *J. Geophys. Res. Atmos.*, 2013, **118**, 11,036, doi: 10.1002/jgrd.50868.
31. J. Pongratz, D. B. Lobell, L. Cao and K. Caldeira, *Nature Clim. Change*, 2012, **2**(2), 101, doi: 10.1038/nclimate1373.
32. L. Xia, A. Robock, J. N. S. Cole, D. Ji, J. C. Moore, A. Jones, B. Kravitz, H. Muri, U. Niemeier, B. Singh, S. Tilmes and S. Watanabe, *J. Geophys. Res. Atmos.*, submitted.
33. A. Robock, *Bull. Atomic Sci.*, 2008, **64**(2), 14, doi: 10.2968/064002006.
34. B. Kravitz, D. G. MacMartin and K. Caldeira, *Geophys. Res. Lett.*, 2012, **39**, L11801, doi: 10.1029/2012GL051652.

35. D. M. Murphy, *Environ. Sci. Technol.*, 2009, **43**(8), 2784, doi: 10.1021/ es802206b.
36. L. M. Mercado, N. Bellouin, S. Sitch, O. Boucher, C. Huntingford, M. Wild and P. M. Cox, *Nature*, 2009, **458**, 1014, doi: 10.1038/nature07949.
37. American Meteorological Society, *Geoengineering the Climate System*, AMS Policy Statement, American Meteorological Society, Boston, MA, 2013; www.ametsoc.org/policy/2013geoengineeringclimate_amsstatement. pdf (accessed 15th January 2014).
38. American Geophysical Union, *Geoengineering the Climate System*, AGU Position Statement, American Geophysical Union, Washington, DC, 2009; www.agu.org/sci_pol/positions/geoengineering.shtml (accessed 15th January 2014).
39. Government Accountability Office, *Climate Engineering: Technical Status, Future Directions, and Potential Responses*, Government Accountability Office, Washington, DC, Report GAO-11-71, p. 135; www.gao.gov/new.items/ d1171.pdf (accessed 15th January 2014).
40. D. W. Keith, E. Parson and M. G. Morgan, *Nature*, 2010, **463**, 426, doi: 10.1038/463426a.
41. G. Betz, *Clim. Change*, 2012, **111**, 473, doi: 10.1007/s10584-011-0207-5.
42. A. Robock, *Peace Security*, 2012, **4**, 226.
43. A. Lin, *Ecol. Law Q.*, in press.
44. J. R. Fleming, *Fixing the Sky: The Checkered History of Weather and Climate Control*, Columbia University Press, New York, 2010.
45. Solar Radiation Management Governance Initiative, *Solar Radiation Management: The Governance of Rsearch*, Royal Society, London, UK, 2011.
46. A. Robock, M. Bunzl, B. Kravitz and G. Stenchikov, *Science*, 2010, **327**, 530, doi: 10.1126/science.1186237.
47. A. Robock, *Ethics, Policy Environ.*, **15**, 202.
48. A. Robock and O. B. Toon, *Sci. Am.*, 2010, **302**, 74.
49. A. Robock and O. B. Toon, *Bull. Atomic Sci.*, 2012, **68**(5), 66, doi: 10.1177/ 0096340212459127.
50. S. M. Gardiner, in *Climate Ethics: Essential Readings*, ed. S. M. Gardiner, S. Caney, D. Jamieson and H. Shue, Oxford University Press, New York, NY, USA, 2010, p. 284.
51. T. Svoboda, K. Keller, M. Goes and N. Tuana, *Public Affairs Q.*, 2011, **25**(3), 157.
52. C. J. Preston, *Engineering the Climate: The Ethics of Solar Radiation Management*, Lexington Books, 2012.
53. Special issue on Geoengineering Research and its Limitations, ed. R. Wood, S. Gardiner and L. Hartzell-Nichols, *Clim. Change, 2013*.
54. S. Rayner, C. Heyward, T. Kruger, N. Pidgeon, C. Redgwell and J. Savulescu, *Clim. Change*, 2013, **121**, 499, doi: 10.1007/s10584-012-0675-2.
55. B. Banerjee, G. Collins, S. Low and J. J. Blackstock, *Scenario Planning for Solar Radiation Management*, Yale Climate and Energy Institute, New Haven, CT, USA, 2013.

56. A. Robock, D. G. MacMartin, R. Duren and M. W. Christensen, *Clim. Change*, 2013, **121**, 445, doi: 10.1007/s10584-013-0777-5.
57. A. Robock, *Clim. Change*, 2011, **105**, 383, doi: 10.1007/s10584-010-0017-1.
58. R. F. Adler, G. J. Huffman, A. Chang, R. Ferraro, P. Xie, J. Janowiak, B. Rudolf, U. Schneider, S. Curtis, D. Bolvin, A. Gruber, J. Susskind, P. Arkin and E. Nelkin, *J. Hydrometrol.*, 2003, **4**, 1147.

Space-Based Geoengineering Solutions

COLIN R. McINNES,* RUSSELL BEWICK AND JOAN PAU SANCHEZ

ABSTRACT

This chapter provides an overview of space-based geoengineering as a tool to modulate solar insolation and offset the impacts of human-driven climate change. A range of schemes are considered including static and orbiting occulting disks and artificial dust clouds at the interior Sun–Earth Lagrange point, the gravitational balance point between the Sun and Earth. It is demonstrated that, in principle, a dust cloud can be gravitationally anchored at the interior Lagrange point to reduce solar insolation and that orbiting disks can provide a uniform reduction of solar insolation with latitude, potentially offsetting the regional impacts of a static disk. While clearly speculative, the investigation of space-based geoengineering schemes provides insights into the long-term prospects for large-scale, active control of solar insolation.

1 Introduction

Industrial civilisation has flourished during a time when the climate has been in a relatively benign and temperate state. This favourable setting has allowed rapid population growth through intensive agriculture, along with prosperity in the West through a mix of fossil fuel energy consumption and innovation-driven economic growth. However, recent concern associated with human-driven climate change has brought into sharp focus that the climate is not static. It is clear that over millennial time-scales the Earth is a dynamic system and the popular view of the climate as being perpetually in equilibrium is only due to the narrow slice of human history through which

---

*Corresponding author

Issues in Environmental Science and Technology, 38
Geoengineering of the Climate System
Edited by R.E. Hester and R.M. Harrison
© The Royal Society of Chemistry 2014
Published by the Royal Society of Chemistry, www.rsc.org

we view the past. Addressing both human-driven, and indeed long-term natural climate change, will be an essential requirement for the sustainability of industrial civilisation and its future prosperity. Human-driven climate change brings these issues to a head.

Geoengineering has risen in prominence as it becomes clear that a reduction in global carbon emissions is highly unlikely in the near-term. This is not unexpected given that GDP per capita in many developing nations is growing sharply, largely driven by fossil fuel energy consumption. While the long-term historical growth of global energy consumption is of the order 2–3% per annum, the historical decarbonisation rate (carbon intensity of primary energy) is of the order 0.2–0.3%.[1] This long-term decarbonisation path has thus far followed a series of energy transitions to fuels of greater energy density and so lower carbon intensity – from wood to coal, then oil, methane and nuclear. However, given the timescale for such energy transitions, a pragmatic view is that absolute global carbon emissions will continue to grow in the decades ahead, even if relative decarbonisation continues through energy innovation. Geoengineering is therefore a timely development which can potentially bridge the gap between future growth in fossil fuel energy consumption and long-term decarbonisation measures.

In this chapter, space-based geoengineering will be considered as a means of reducing solar insolation to counteract the increased radiative forcing from carbon dioxide emissions. In section 2 a review of space-based geoengineering is provided, followed by a detailed analysis of several schemes. Section 3 considers the use of solid occulting disks in equilibrium sunward of the classical L_1 equilibrium point between the Sun and Earth, section 4 considers an L_1 dust cloud, while section 5 considers the use of an orbiting disk in the vicinity of L_1 whose phasing is selected to provide a uniform reduction in insolation with latitude on average. While clearly speculative, the investigation of space-based geoengineering schemes puts bounds on the design space of the geoengineering problem and provides an insight into the long-term prospects for active control of solar insolation.

2 Space-based Geoengineering

Geoengineering is the process of measured, active intervention to regulate the climate in a controlled manner with presumed beneficial effect. While geoengineering has a long history in various guises, as reviewed by Keith,[2] it is only recently that geoengineering has been seriously considered for deployment. As global carbon emissions continue to grow, it is generally agreed that it is now prudent to pursue an active programme of geoengineering research, including associated technical and regulatory oversight.

In order to mitigate climate change impacts, a range of geoengineering measures have been proposed. These measures attempt to offset enhanced radiative forcing due to increased carbon dioxide loading in the atmosphere and so regulate the global mean surface temperature. Proposed measures include solar radiation management (SRM) through the deposition of sulfur aerosols in the stratosphere[3,4] and cloud whitening,[5,6] along with ocean fertilisation[7] and biochar[8] to accelerate carbon dioxide removal (CDR) processes. Ambitious space-based methods to directly scatter solar radiation have also been proposed,[9,10] as will be discussed in this chapter. The effectiveness of these geoengineering measures has been considered and initial trade-offs performed.[11] Studies indicate that some near-term schemes, such as sulfur aerosol deposition, are likely to be effective and incur relatively modest costs, in comparison to rapid decarbonisation. Although critics of geoengineering have argued that coarse modification of solar insolation will lead to a range of uncertain and undesirable consequences, it has been shown that impacts are potentially modest compared to the direct effects of increased radiative forcing.[12]

Aside from aerosol deposition and carbon sequestration, large-scale geoengineering using orbiting reflectors has long been considered by various authors to manipulate solar insolation.[9,13–18] These concepts centre on fabricating and deploying a large occulting disk (or many smaller disks) to reduce the total solar insolation in order to mitigate increased radiative forcing by carbon dioxide, shown schematically in Figure 1. For example, the

Figure 1   Assembly of a large occulting disk from a swarm of discrete elements as a possible route to space-based Geoengineering (adapted from Dario Izzo, European Space Agency/Advanced Concepts Team).

use of vast numbers ($\sim 5 \times 10^4$) of 100 km^2, actively controlled occulting disks in Earth orbit has been considered, but would likely lead to an apparent flickering of the Sun ($\sim 2\%$ amplitude) and would create a significant orbital debris hazard.[3] In addition, various proposals for an artificial ring of passive scattering particles in Earth orbit have been documented with a mass of order 2×10^9 tonnes.[19,20] A recent addition to this class of concepts is the use of clouds of dust grains located at the stable Earth–Moon triangular Lagrange points L_4 and L_5.[21] In this scheme, in the order of 2×10^{11} tonnes of lunar or cometary dust is deposited in the Earth–Moon system forming clouds at L_4 and L_5. As with other occulting schemes, the dust would lead to a reduction in solar insolation to offset radiative forcing by carbon dioxide. However, since each cloud only reduces solar insolation for a relatively short period each month, when the cloud is between the Earth and the Sun, significant mass is required to ensure a large optical depth and so useful cooling on average.

A perhaps more effective, although still clearly ambitious scheme, is to deploy a large occulting disk (or disks), typically with a total mass of order 10^7–10^8 tonnes, close to the L_1 Lagrange equilibrium point on the Sun-Earth line, some 1.5×10^6 km sunward of the Earth. At this point the gravitational force of the Sun is balanced by the gravitational force of the Earth and the centripetal force due to orbital motion about the Sun. The equilibrium location at the L_1 point is unstable, necessitating the use of active control. Indeed, the use of a large number of smaller occulting disks would mitigate the potentially catastrophic effect of the loss a single large disk.[18]

The use of L_1 for geoengineering has been revisited by Angel who proposes swarms of engineered thin film refractive (rather than reflective) disks with a total mass of order 2×10^7 tonnes.[9] Rather than directly reflecting solar radiation, the refracting disks scatter sunlight, but are highly engineered thin film devices which require terrestrial fabrication and launch at extremely high cost. However, it is possible that much simpler partly reflecting disks could be fabricated *in situ* from near Earth asteroid resources.[18] More recent concepts for space-based geoengineering will be considered later in this chapter.

As has been noted elsewhere,[19] a key advantage of using large solar reflectors for geoengineering is the vast energy leverage obtained in a relatively short duration. The total accumulated solar energy intercepted by the reflector quickly grows beyond the energy required for its fabrication, leading to a highly efficient tool for climate engineering. While solar reflectors offer many advantages, there are clearly significant challenges associated with the fabrication and active control of such large, gossamer structures. Again, it is almost certain that such structures would be fabricated in-orbit, either using lunar material or material processed from a suitable near Earth asteroid. Therefore, a prerequisite for space-based geoengineering is the long-term capability to effectively and economically exploit the resources of the moon or asteroids.

3   Lagrange Point Occulting Disks

3.1   Occulting Solar Disks

The concept of using a large occulting disk (or disks) near the Sun-Earth L_1 equilibrium point to reduce solar insolation has been discussed by various authors, as noted in section 2. In this section it will be shown that there is in fact a minimum system mass which can be obtained if the disk is positioned at an optimum location along the Sun–Earth line, sunward of the classical L_1 point. This optimum location is found from an analysis of the three-body dynamics of the problem, with the addition of solar radiation pressure on the disk.[22] Solar radiation pressure is due to transfer of momentum transported by solar photons reflecting off the disk. The location of the disk can be optimised since the solar radiation pressure exerted on the disk will generate a new equilibrium position, sunward of the classical L_1 point. If the disk mass is reduced, the solar radiation pressure induced acceleration will increase and so the equilibrium point will be displaced further sunward of the classical L_1 point.[17] However, as the disk is displaced sunward, the required disk area to maintain the necessary reduction in solar insolation at the Earth will grow, leading to an increase in disk mass. These two competing processes must then be balanced in order to minimise the total disk mass through an optimum choice of disk location.

3.2   Occulter Orbit

For a disk of radius R_S at some distance r_S from the Earth, the disk will subtend a solid angle Ω_S of $\pi R_S^2/r_S^2$, as shown in Figure 2. Similarly the Sun, of radius R_O at distance r_O from the Earth, will subtend a solid angle Ω_O of $\pi R_O^2/r_O^2$, so that the disk will partially occult the Sun and reduce the solar insolation at the Earth by a factor Ω_S/Ω_O. By partly occulting the solar disk, the solar flux F is reduced such that $\delta Q = \delta F/4$, however the relative change in insolation $\delta Q/Q$ is identical to the relative change in flux $\delta F/F$. Therefore, the reduction in insolation produced by the occulting disk is defined as follows:

$$\frac{\delta Q}{Q} = \left[\frac{R_S}{R_O}\right]^2 \left[\frac{r_O}{r_S}\right]^2 \qquad (1)$$

Using eqn (1), the required disk radius may now be obtained as a function of its distance from the Earth r_S to provide the required change in solar insolation of $\delta Q/Q$ as follows:

$$R_S = R_O \left[\frac{r_S}{r_O}\right] \left[\frac{\delta Q}{Q}\right]^{1/2} \qquad (2)$$

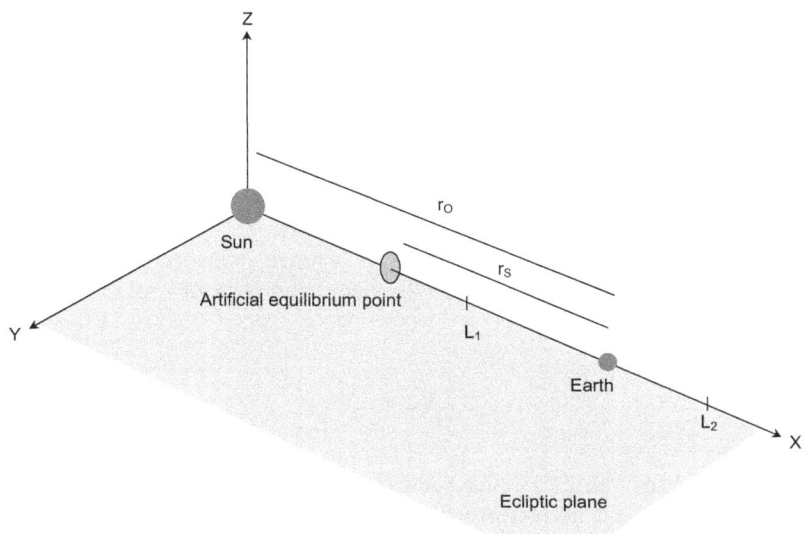

Figure 2 Occulting solar disk stationed along the Sun–Earth line at an artificial equilibrium point. L_1 and L_2 are classical 3-body equilibrium points.

However, before the occulting disk is sized, the optimum distance of the disk from the Earth can be determined to minimise the total disk mass. As discussed in section 3.1, the disk will be located near the Sun–Earth L_1 equilibrium point. However, due to the solar radiation pressure acting on the disk, the equilibrium point will be displaced sunward of the classical L_1 point. A trade-off therefore exists between lowering the disk mass and displacing the equilibrium point sunward, and ultimately increasing the disk mass due to the increased disk area required to provide partial occultation of the Sun. This trade-off leads to an optimum disk location which will minimise the total mass of the system.

The condition for equilibrium of the occulting disk in the Sun–Earth three-body problem can be determined from a simple force balance. The eccentricity of the Earth's orbit is neglected as is the lunar gravitational perturbation. Although the general solution for artificial equilibria for a reflector is known,[22] only a simple 1-dimensional problem need be considered here to locate the displaced equilibrium point. Similarly, it will be assumed that the disk station-keeps close to the Sun–Earth line to provide direct shading. Since the mass of the Earth M_E is essentially negligible relative to the solar mass M_O, the centre-of-mass of the Sun–Earth system will be taken as being located at the centre-of-mass of the Sun, as shown in Figure 2. This approximation has a negligible effect on the subsequent analysis. The condition for equilibrium may now be obtained by balancing the gravitational force from the Sun and the Earth, the centripetal force and the solar radiation pressure induced acceleration experienced by the occulting disk a_S such that:

$$\frac{GM_E}{r_S^2} - \frac{GM_O}{(r_O - r_S)^2} + \omega^2 (r_O - r_S) + a_S = 0, \quad \omega = \sqrt{\frac{GM_O}{r_O^3}} \quad (3)$$

where ω is the orbital angular velocity of the Earth relative to the Sun and G is the gravitational constant. The inverse square solar radiation pressure induced acceleration experienced by the occulting disk of mass M_S and area A_S may be written as:

$$a_S = \frac{2\kappa P_E A_S}{M_S} \left[\frac{r_O}{r_O - r_S} \right]^2 \quad (4)$$

where P_E (4.56×10^{-6} Nm^{-2}) is the solar radiation pressure experienced by an absorbing surface at 1 astronomical unit (r_O) and κ is a function of the optical properties of the disk. It can be shown that for a specular reflector with Lambertian thermal re-emission the function κ is given by:[23]

$$\kappa = \frac{1}{2} \left[(1 + \eta) + \frac{2}{3} (1 - \eta) \frac{\varepsilon_F - \varepsilon_B}{\varepsilon_F + \varepsilon_B} \right] \quad (5)$$

where η is the specular reflectivity of the disk, ε_F is the emissivity of the front (Sun facing) side of the disk and ε_B is the emissivity of the rear (Earth facing) side of the occulting disk, while the disk has an area A_S of πR_s^2. Using eqn (4) and (5), the disk mass M_S may now be written as:

$$M_S(r_S) = 2\pi\kappa P_E R_O^2 \left[\frac{\delta Q}{Q} \right] \left[\frac{r_S}{r_O - r_S} \right]^2 \frac{1}{a_S(r_S)} \quad (6)$$

where a_S is determined from eqn (3). Since eqn (6) is now a function of r_S only, the variation of the mass of the occulting disk with location along the Sun–Earth line can be investigated to attempt to minimise the disk mass.

3.3   Occulter Sizing

The mass of the occulting disk may now be determined for a required reduction in solar insolation. For a fixed disk area, changing the disk mass will alter the solar radiation pressure acceleration experienced by the disk and so will influence the location of the equilibrium point. Assuming an insolation reduction $\delta Q/Q$ of 1.7%,[10] the variation of the disk mass with equilibrium point location is shown in Figure 3, where it can be seen that there are two limiting conditions. First, as the disk is located closer to the classical L_1 equilibrium point, 1.5×10^6 km sunward of the Earth, the disk mass grows and

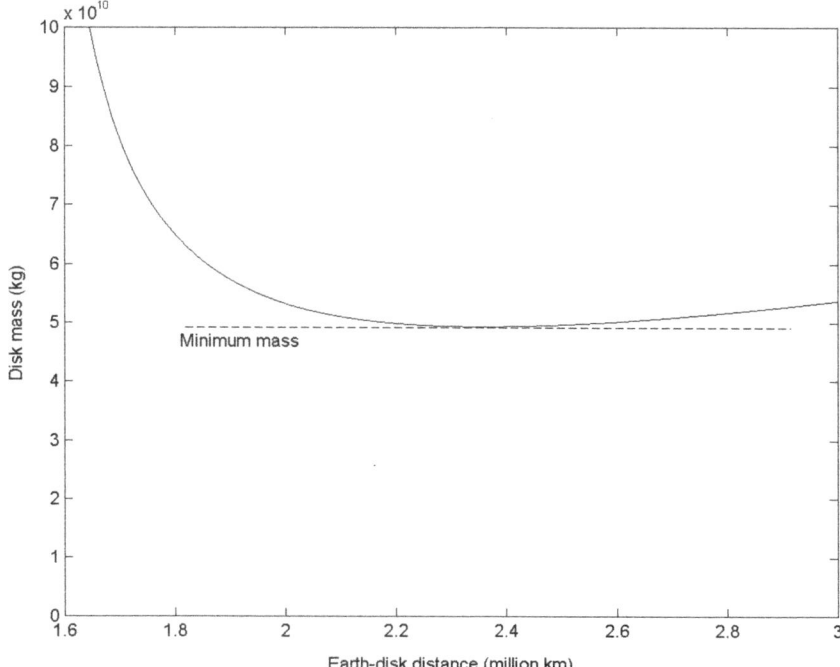

Figure 3 Optimum occulting disk location ($\kappa = 0.17$). The classical L_1 point is at a distance of 1.5×10^6 km from the Earth.

is unbound as the classical Lagrange point is approached. This growth in mass is required to reduce the solar radiation pressure induced acceleration experienced by the disk, which would otherwise displace the equilibrium point sunwards. Similarly, as the location of the disk is displaced sunwards, the required mass of the disk will fall due to the increased solar radiation pressure induced acceleration required for equilibrium. However, as the disk is displaced sunwards significantly from the classical L_1 point, its mass will start to grow as the disk area increases to maintain the required solid angle subtended at the Earth to reduce the total solar insolation. These two competing processes lead to a minimum disk mass, as can be seen in Figure 3.

The minimum disk mass can now be determined by finding the turning point of eqn (6). It can be seen that there is a single location which will minimise the disk mass, independent of the required reduction in solar insolation or the disk optical properties. This location can be found by minimising the function:

$$f(r_S) = \left[\frac{r_S}{r_S - r_O}\right]^2 \frac{1}{a_S(r_S)} \tag{7}$$

Table 1 Occulting disk optical properties.

Disk type	η	ε_F	ε_B	κ
A (reflecting)	0.82	0.06	0.06	0.91
B (non-reflecting)	0.00	0.01	0.50	0.17

Table 2 System level trade-off.

Geoengineering concept	Mass (tonne)	Area (km^2)	Areal density (gm^{-2})
Struck (Lunar $L_{4/5}$ dust cloud, ref. 21)	2.1×10^{11}	–	–
Pearson (Earth orbit dust ring, ref. 20)	2.3×10^9	1.1×10^8	–
McInnes A (Solar L_1 reflecting disks, ref. 17,18)	2.6×10^8	6.57×10^6	40.2
McInnes B (Solar L_1 absorbing disks, ref. 17,18)	5.2×10^7	6.57×10^6	7.9
Angel (Solar L_1 refracting disks, ref. 9)	2.0×10^7	4.70×10^6	4.2

Table 3 Mass comparison with terrestrial engineering ventures.

Mass scale	Mass (tonne)	Engineering venture
10^5	6.5×10^5	'Knock Nevis' oil tanker (fully laden)
10^6	6×10^6	Great pyramid of Giza
10^7	6×10^7	Concrete used for Three Gorges dam
10^8	2×10^8	Water stored in London's reservoirs
10^9	7×10^9	World annual CO_2 emissions

where it is found that $f'(r_s) = 0$ when the disk location r_s is 2.36×10^6 km sunwards of the Earth. This optimum location is sunward of the classical L_1 point at 1.50×10^6 km and again is independent of the disk properties, representing the true optimum location for an occulting disk (or disks). Assuming that the disk must provide a reduction in solar insolation $\delta Q/Q$ of 1.7%, a disk with an effective radius of 1450 km (or an equivalent area from a large number of smaller disks) and a total mass of 2.6×10^8 tonne is required if $\kappa \sim 0.91$, representative of a reflecting metallic occulting disk.

However, if $\eta \sim 0$ and $\varepsilon_F \sim 0$ then $\kappa \sim 0.17$, representative of a non-reflecting black occulting disk, substantial mass savings occur, although the optimum location of the disk remains unchanged. In this case a total mass of 5.2×10^7 tonne is required, again with an effective radius of 1450 km, as detailed in Tables 1 and 2. A non-reflecting disk could in principle be fabricated using a thin layer of carbon, vacuum deposited on a metal film. For *in situ* manufacturing using a small near Earth asteroid a carbon resource is readily available. While the mass and scale of the disk is clearly vast, for comparison the masses of a range of terrestrial engineering ventures are listed in Table 3. It can be seen that the Chinese Three Gorges Dam requires approximately

6×10^7 tonne of concrete, and so forms a structure with a comparable mass to the occulting disk (or disks). While the challenges posed by space-based geoengineering are clearly significant, it interesting to note that measured in terms of mass, such large-scale geoengineering represents a venture of comparable scale to current large-scale terrestrial engineering ventures.

4   Lagrange Point Dust Cloud

While space-based geoengineering is in principle an effective method of solar insolation reduction, it is clearly prohibitively expensive using current technologies.[24] This is primarily due to the need to manufacture, launch and operate such large structures in space. For example, the refracting disk scheme proposed by Angel requires the launch of 2×10^7 tonnes of material from Earth.[9] Whilst the *in situ* manufacture of material can reduce the engineering challenge, and associated costs, it would clearly be extremely difficult to implement.

A new method to attempt to reduce the engineering complexity of space-based geoengineering is to remove the need to manufacture and launch significant quantities of material from Earth. This can in principle be achieved by the utilisation of the population of small near-Earth asteroids (NEAs). This section will describe several methods through which *in situ* asteroid material can be captured and subsequently used to form large clouds of dust to reduce solar insolation. The use of un-processed dust is in contrast to the use of highly engineered reflecting or refracting disks.

As discussed in section 2, several previous studies have proposed the use of large clouds of dust as a means of space-based geoengineering, with clouds located at the L_4 and L_5 points of the Earth–Moon system[21] and also as a diffuse Earth ring.[20] These methods have the disadvantage of creating highly variable insolation reduction over the surface of the Earth, and for the Earth–Moon Lagrange point case insolation reduction which is strongly modulated by the motion of the Moon. In contrast, a dust cloud deployed at the L_1 equilibrium point on the Sun–Earth line will largely eliminate this problem by creating a constant insolation reduction with a relatively modest variability over the Earth's surface. However, since a cloud located at the Sun–Earth L_1 point will be unstable the dust cloud will therefore dissipate over time and will require re-plenishment. Two different dust cloud methods will now be discussed; a dissipating cloud released in the vicinity of the L_1 point and a cloud gravitationally anchored by a captured near Earth asteroid, as shown schematically in Figure 4.

4.1   *Dissipating Dust Cloud*

The dissipating cloud will be modelled by the circular restricted three-body problem, where only the Sun and Earth influence the dynamics of the dust

Figure 4   Sun–Earth L_1 Lagrange point dust cloud generated *in situ* from near Earth
asteroid material.

particles. The dimensionless equations of motion of a dust particle in a
frame of reference rotating with the Earth are given by:[25]

$$\ddot{x} - 2\dot{y} = \frac{\partial U}{\partial x}$$

$$\ddot{y} + 2\dot{x} = \frac{\partial U}{\partial y} \tag{8}$$

$$\ddot{z} = \frac{\partial U}{\partial z}$$

where the non-dimensional 3-body potential function U is given by:

$$U(x,y,z) = \frac{1}{2}\left(x^2 + y^2\right) + \frac{1-\mu}{\rho_1(x,y,z)} + \frac{\mu}{\rho_2(x,y,z)} \tag{9}$$

and the coordinates of the dust particle are defined in the Cartesian frame
shown in Figure 2. Here the mass ratio of the Earth to the total system mass
is $\mu = M_E/(M_S + M_E)$, while $\rho_{1,2}$ are the distances of the dust particle from the
Sun and Earth respectively, normalised to 1 astronomical unit (AU), the
distance between the Sun and the Earth. In dimensionless co-ordinates
the Sun of mass M_S and Earth of mass M_E are positioned at $(-\mu,0,0)$ and
$(1-\mu,0,0)$, respectively.

Using these equations of motion, the dynamics of a dust particle can be
described by propagating its initial position using a state transition matrix. The

state transition matrix propagates a set of initial conditions forwards in time and is generated by integrating the linearised dynamics of the circular restricted three-body problem in the vicinity of the L_1 point, as detailed in ref. 25.

4.1.1 Solar Radiation Pressure. The effect of solar radiation pressure has been described earlier in section 3.2 and can be quantified using the 'lightness' parameter, β, which is defined as the ratio of the force due to solar radiation pressure and solar gravity as follows:

$$\beta = \left| \frac{F_{rad}}{F_g} \right| \approx 570 \frac{Q}{\rho R_{gr}} \qquad (10)$$

where ρ (kg m^{-3}) is the grain density and R_{gr} (μm) is the radius of the dust grain, and Q characterises momentum transfer from solar photons to the dust grain.[25] Since both solar radiation pressure and solar gravity have an inverse square variation with distance from the Sun, this is a dimensionless parameter.

For relatively large particles, $R_{gr} > 1$ μm, Q varies little but as the grain size decreases the interaction between the solar photons and the dust grains becomes more complex. The expected value of β for a range of particle radii can be determined using Mie scattering theory for different composition models.[26] The results for a typical asteroid dust grain can be seen in Figure 4. This shows that β peaks with a value of approximately 0.9 at a radius of 0.2 μm before decreasing to 0.1 for a radius of 0.01 μm. Also shown in Figure 5 are vertical lines defining the grain sizes that will be modelled later, related to their corresponding values of β, as listed in Table 4.

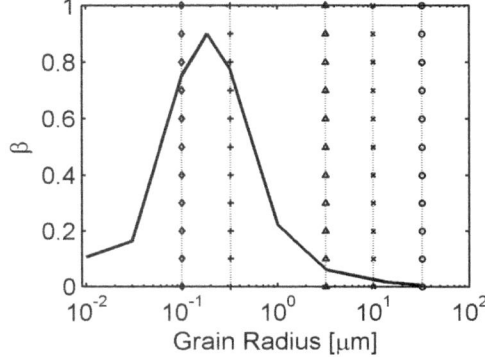

Figure 5 Variation in the lightness parameter, β, (ratio of the force due to solar radiation pressure and solar gravity) with particle radius for an asteroid dust grain model as described in ref. 26. Vertical lines correspond to the mass requirement results shown in Figure 6.

Table 4   Grain radii simulated and the corresponding values
of β and displacement of the equilibrium position
with respect to the classical L_1 point.

Grain Radius [µm]	β	Displacement [km]
32	0.005	2500
10	0.018	9000
3.2	0.061	32 000
0.32	0.772	950 000
0.1	0.751	875 000

Due to the inverse square scaling of solar radiation pressure with distance
from the Sun, its effect is to reduce the effective gravitational potential of the
Sun. Hence, the mass parameter μ for the three-body problem must now
defined as:

$$\mu = \frac{M_E}{(1 - \beta)M_S + M_E} \tag{11}$$

Due to the increase in the value of μ with increased β, the L_1 equilibrium
point is found to be displaced towards the Sun. For example, a dust grain
with $\beta = 0.772$ will experience a sunward shift of 950 000 km in order to
remain in equilibrium.

4.1.2 Dust Cloud Attenuation. The attenuation of the Lagrange point
dust cloud can be determined using the Beer–Lambert law defined as:

$$I = I_0 e^{-\sigma_{gr} \int \rho_n(l) dl} \tag{12}$$

where I and I_0 are the intensity of the attenuated and incident light on the
dust cloud, l is the path length through the cloud, σ_{gr} is the physical cross-
section of a single dust grain and ρ_n is the number density of dust particles
at a given point along the path. In this analysis several simplifying assumptions are made. First, the dust cloud is assumed to be spherical and is
comprised of a single dust grain size with a homogeneous number density.
Then, the dust cloud is assumed to be initially static before being propa-
gated. Therefore, the attenuation of the solar flux at a given point is
dependent upon a single variable, the number density of particles.

As noted earlier, the state transition matrix ϕ can be used to propagate the
dynamics of an ensemble of dust particles. The number density ρ of particles
at a given point with respect to their initial number density can then be
found using:

$$\rho(\mathbf{x}\,;\,t) = \left\| \frac{\partial \mathbf{v}(t)}{\partial \mathbf{v}(0)} \right\| H(r_{cloud} - \| \phi^{-t}(\mathbf{x}, \mathbf{v}_*)_r \|) \tag{13}$$

where the first term determines the relative density of particles at
position some $\mathbf{x}$ and velocity $\mathbf{v}$, whilst the second term is the Heaviside

function to determine whether the initial position was within the initial cloud radius, r_{cloud}. Equation (13) can therefore be used to propagate the dynamics of the dust cloud, which can then be used in the Beer–Lambert integration along paths between multiple points on the surfaces of the Sun and Earth to determine the average insolation reduction generated by the dust cloud.

4.1.3 Insolation Reduction. The key design parameter for the dust cloud methods of space-based geoengineering is the mass of material required per year to achieve the 1.7% solar insolation reduction discussed in section 3.3. First, a spherical dust cloud will be considered, released from either the displaced L_1 equilibrium position (due to solar radiation pressure) or the classical L_1 point for a range of dust grain sizes and initial cloud radii. The Beer–Lambert law can then be used to calculate the reduction in solar insolation experienced for a given initial number density of particles. From this the mass of material required to generate the necessary insolation reduction can be calculated. The results for a range of grain sizes for clouds ejected at the L_1 point can be seen in Figure 6.

In general it would be expected that the larger particles, which have smaller values of β, would require less mass since they have a greater average lifetime due to the lower effect of solar radiation pressure. However, this is not the case as the decrease in grain size provides a greater mass saving than the longer lifetime of the larger particles, with the optimum solution occurring for the smallest grain radius of 0.1 μm. For an optimum cloud radius of 4000 km the mass requirement is 7.6×10^{10} kg yr^{-1}. In comparison to a dust cloud at

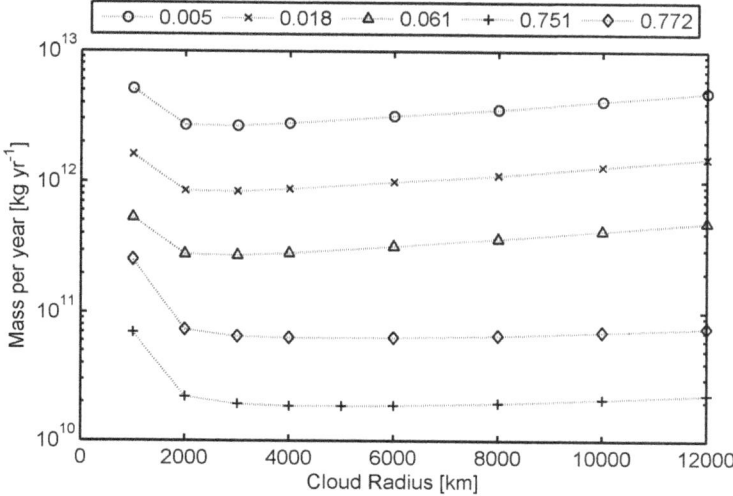

Figure 6   Mass requirement of dust for a steady state solution of clouds ejected at the L_1 point for varying initial cloud radii for the five grain β-values considered.

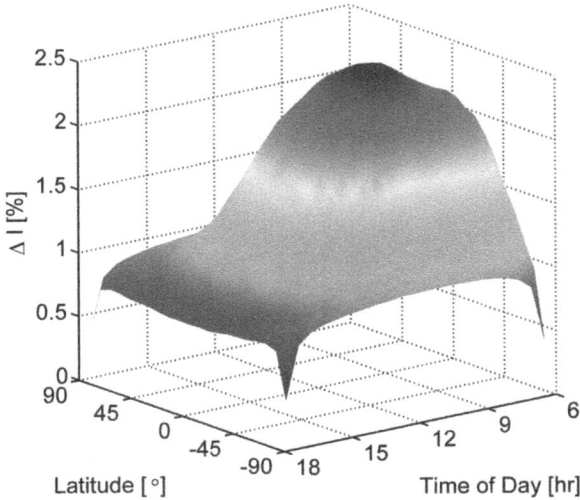

Figure 7 Insolation change over the surface of the Earth for the steady state solution with an initial cloud of radius 4000 km and grain size of 0.1 μm
released at the classical L_1 point.

the Earth–Moon L_4/L_5 points, this represents a mass saving of several orders
of magnitude.[21]

A map showing the insolation change over the Earth's surface for a cloud
of radius 4000 km and grain size 0.1 μm released at the L_1 point can be seen
in Figure 7, where the tilt of the Earth's axis is not taken into account. It can
be seen that the insolation change is greatest towards one side of the Earth
due to the drift of the cloud away from the initial release position being
biased in one direction due to the three-body orbital dynamics. This will lead
to greater shading in the 'morning' region of the Earth.

4.2   Anchored Dust Cloud

The dust cloud method described in section 4.1 requires the constant re-
plenishment of significant amounts of material. To reduce this replenishment requirement the possibility of gravitationally anchoring the dust cloud
using a captured near-Earth asteroid can be considered. This anchoring
scheme will be achieved by placing a captured body at the Sun–Earth L_1
point, therefore creating an artificial four-body problem. The size of cloud
that can be anchored at the L_1 point will then be estimated and the potential
for insolation reduction determined by the use of a solar radiation model as
described in other work by the authors.[25] Clearly this is a highly speculative
scheme, but again places bounds of the deign space of the problem.

4.2.1 Four-body Problem.   The dimensionless equations of motion of a
dust grain in the circular restricted Sun–Earth three-body problem have

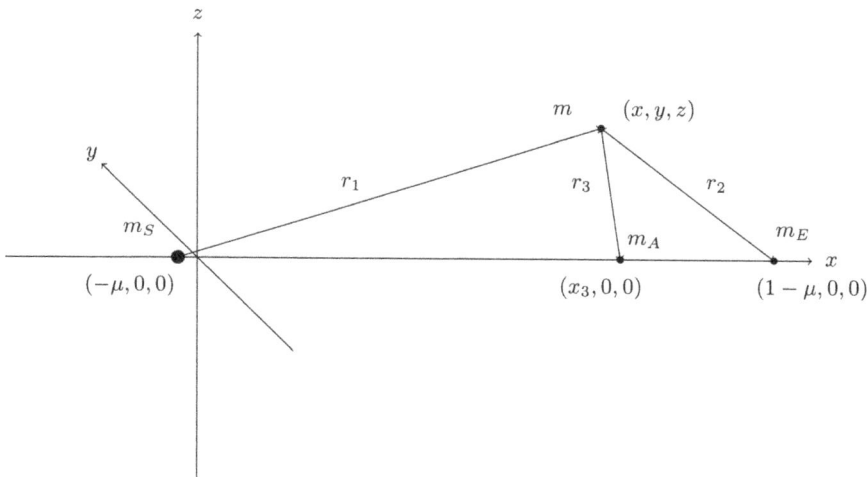

Figure 8 Four-body problem with Sun (M_S), Earth (M_E), asteroid (M_A) and dust particle (m).

been discussed previously in section 4.1. These equations can be modified to define a circular restricted four-body problem which includes a small asteroid captured at L_1, so that the new effective potential is defined by;

$$U(x,y,z) = \frac{1}{2}\left(x^2 + y^2\right) + \frac{1-\mu}{\rho_1(x,y,z)} + \frac{\mu}{\rho_2(x,y,z)} + \frac{\gamma}{\rho_3(x,y,z)} \qquad (14)$$

The parameter γ is the mass fraction of the asteroid M_A in relation to the mass of the three-body system, $\gamma = M_A/(M_S + M_E)$, with the additional scalar distance ρ_3 being the separation between the asteroid and dust particle respectively, as shown in Figure 8.

Due to the small dust grain sizes, the effect of solar radiation pressure is again included, having the effect of reducing the effective solar gravitational potential by a factor $(1-\beta)$. As before this has the effect shifting the L_1 equilibrium position sunwards along the Sun–Earth line. When the gravitational potential of a body placed at the classical L_1 point is considered, two new equilibrium positions appear which are located on the Sun–Earth line, either side of the classical L_1 position, as shown in Figure 9 and Figure 10. These new equilibria, like the conventional L_1 position, are unstable, but are bound to the asteroid, thus approximating the size of the gravitationally anchored dust cloud.

4.2.2 Zero Velocity Curve. The speed V of a particle in the circular restricted four-body problem can be described using the Jacobi integral as;

$$V^2 = 2U(x,y,z) - C \qquad (15)$$

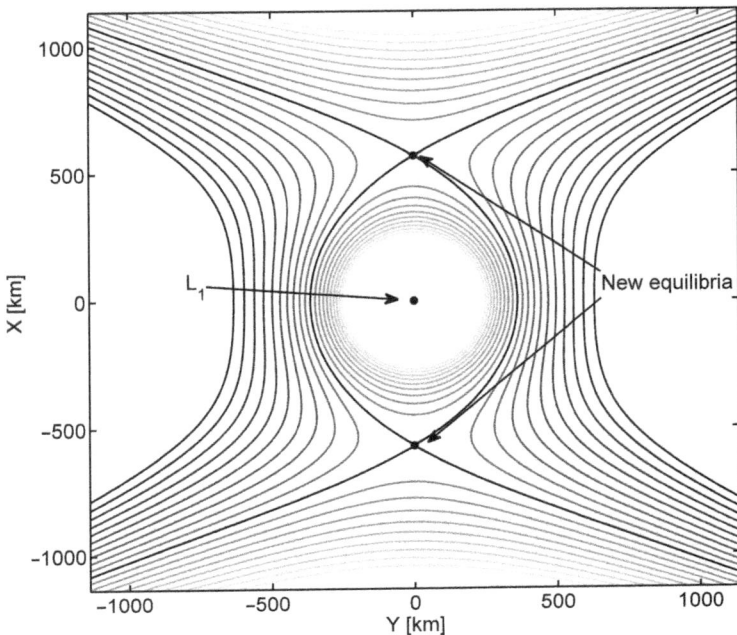

Figure 9   Effective potential of the four-body problem for a body of mass 1×10^{15} kg captured at the classical L_1 point for $\beta = 0$, with the bold line showing the contour with a Jacobi constant of the equilibrium point that encloses the asteroid.

where C is the Jacobi constant. The Jacobi integral is a conserved quantity and defines the volume of space within which dust grains can move. Since kinetic energy can only be strictly positive, it follows from eqn (15) that the particle can only move within a region delimited by a zero velocity curve (*i.e.*, when the right hand side of eqn (15) vanishes). This constraint can be used to investigate the size of the region around the asteroid at L_1 where a particle can become trapped if the energy, or Jacobi constant, of the particle is not large enough for escape. It is assumed that within this region particle dust motion is collisionless. Clearly, the maximum enclosed volume will be found for a zero velocity surface with a Jacobi constant equal to that of one of the new equilibrium points in the circular restricted four-body problem. The corresponding Jacobi constant can then be found by combining eqn (14) and eqn (15) and noting that the equilibrium points lie along the Sun–Earth line. The surfaces that are defined by this analysis can then be found and the shape and volume fully enclosed by the surfaces determined. Examples of these surfaces can be seen in Figure 9 and Figure 10 for an asteroid with a mass of 1×10^{15} kg placed at the classical L_1 position for $\beta = 0$ and $\beta = 0.001$, respectively. It can be seen that for even small values of β the shape of the zero-velocity curve becomes distorted and shrinks in size. However, if an asteroid is placed at the new equilibrium position, found when the effect of solar radiation pressure is included, the increase in β does not significantly

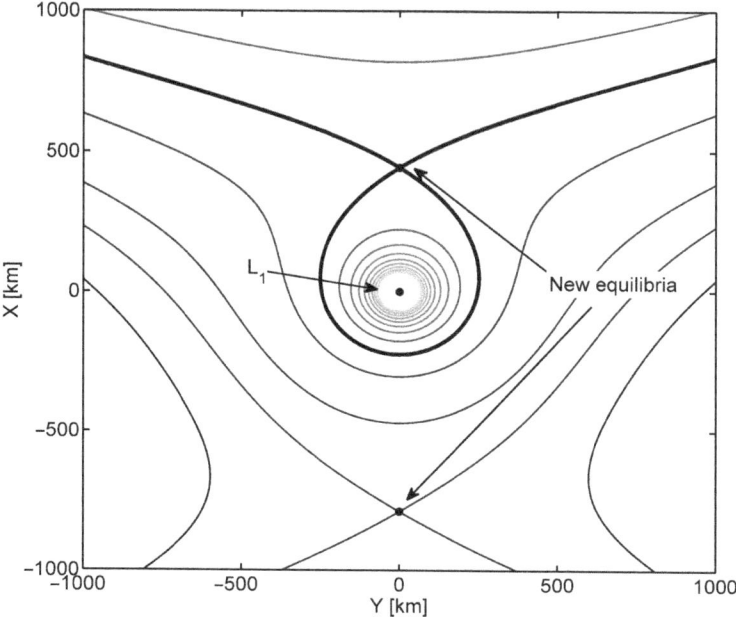

Figure 10 Effective potential of the four-body problem for a body of mass 1×10^{15} kg captured at the classical L_1 point for $\beta = 0.001$, with the bold line showing the contour with a Jacobi constant of the equilibrium point that encloses the asteroid.

affect the size of the bound region. It can therefore be concluded that an asteroid should be positioned at the displaced equilibria to deliver the maximum reduction in solar insolation.

4.2.3 Effect on Solar Insolation. The effect that the dust cloud, confined by the zero velocity surface, has on solar insolation has been determined for a set of real asteroids. The set was chosen by calculating the approximate change in speed, or Δv, required to capture the known near-Earth asteroids with a mass larger than 1×10^{13} kg (up to a mass of 1.3×10^{17} kg for the largest known near Earth asteroid Ganymed 1036). Then a Pareto front can be generated showing impulse ($I = M_A\,\Delta v$) against asteroid mass to select the most efficient asteroids to capture. The masses of these asteroids can then be used during the calculation of solar insolation reduction, for both the classical and displaced equilibrium positions for a dust grain size of 32 μm, representative of real material.

The insolation reduction can be computed using a numerical solar radiation model developed by the authors.[25] Since the maximum achievable insolation reduction is clearly required, the model initially assumes that all light passing through the zero velocity curve is blocked. These results, shown in Figure 11, demonstrate a linear trend on a log–log plot of insolation reduction and captured asteroid mass, with the maximum insolation

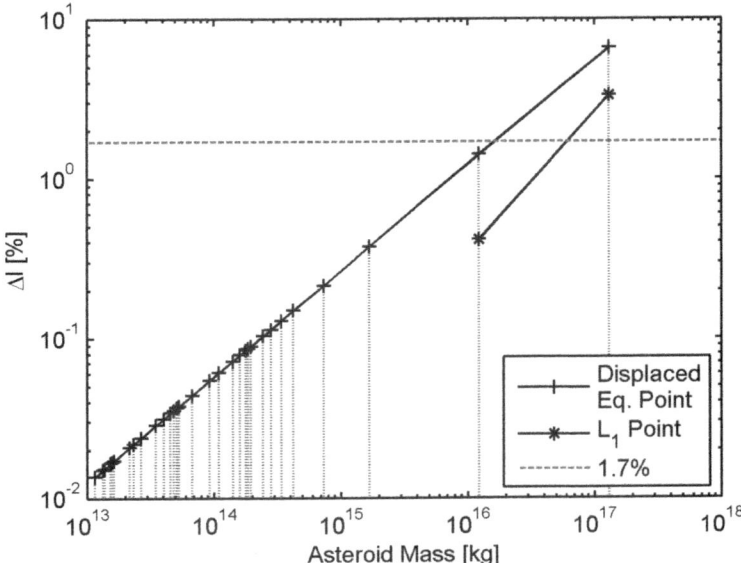

Figure 11 Maximum insolation change available for asteroids on the Pareto front situated at the displaced equilibrium position and the classical L_1 point with labels for the three largest bodies.

reduction of 6.58% being achieved for the asteroid Ganymed at the displaced equilibrium position, and with a maximum insolation reduction of 3.3% for the classical L_1 position. The maximum change in solar insolation reduces significantly for the next largest asteroid, to 1.42% and 0.42% for the displaced equilibrium and classical L_1 positions respectively. These values do not meet the required 1.7% insolation requirement for geoengineering, however, insolation reduction achieved by utilising the next few asteroids could still provide a significant reduction in solar insolation.

To achieve an insolation change of 1.7%, a homogeneous number density of dust particles of 120 m^{-3} is required within the zero velocity curve at the L_1 position, giving a mean free path of approximately 10 000 km. The initial velocity from the surface of a body such as Ganymed, assuming a radius of 32 km, can then be calculated using eqn (13) to be 23 m s^{-1}. This results in a collision timescale of 5.3 days, confirming that the motion of the dust particles can be assumed to be largely collisionless.

5   Optimal Configuration for Lagrange Point Occulting Disks

5.1   GREB Climate Models

A common procedure when attempting to size space-based geoengineering schemes is to size the reflector (as in section 3), or the quantity of dust (as in section 4) needed to reduce solar insolation by some required amount. As discussed in section 2, a suitable reduction in the solar insolation has been shown to yield a decrease of the global mean surface temperature sufficient to compensate for the expected human-driven increase in global mean temperature.[27] However, it has been demonstrated that in such a geoengineered world there would still be significant changes to regional climates, such as warming at high latitudes and cooling below the pre-industrial climate at the tropics.[12]

In order to finesse space-based geoengineering concepts, this section makes use of a Globally Resolved Energy Balance Model (GREB) to provide insights into the coupling between an occulting disk and the Earth's climate. GREB represents a mid-point between oversimplified zero and one-dimensional energy balance models and the complexity of Coupled General Circulation Models (CGCM), while it has been shown to capture the main characteristics of human-driven climate change.[28] GREB therefore provides a simple conceptual model that is on a comparable horizontal grid resolution to CGCM simulations, but provides an efficient tool to simulate approximate climate responses to changes in solar insolation.

Figure 12 represents the climate response after 50 years to a doubling of CO_2 concentration as computed by GREB, assuming a step change in CO_2 concentration from 340 ppm (level similar to that of 1980s) to a concentration of 680 ppm. The GREB response to this scenario matches reasonably well the climate predictions from much more complex models,[28] taking into account that a more realistic scenario should consider a continuous rather than step increase in CO_2 concentration, as defined by possible scenarios issued by the Intergovernmental Panel on Climate Change (IPCC). Nevertheless this case provides a useful baseline scenario to measure the impact of human-driven climate change. This section will now attempt to offset the impact of the step change in CO_2 concentration by the optimal placement of an occulting disk near the Sun–Earth L_1 equilibrium point. In contrast to

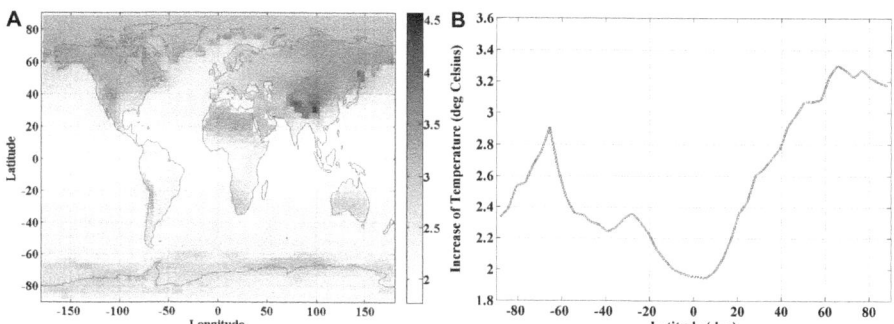

Figure 12 $2 \times CO_2$ world scenario as computed by the GREB model. Initial CO_2 concentration set to 340 ppm, thus the doubled concentration is 680 ppm. A) Average change of annual mean surface temperature, computed by comparing the annual mean surface temperature of the $2 \times CO_2$ case with the results of $1 \times CO_2$. B) Annual mean latitudinal increase of temperature. The averaging of the mean latitudinal temperature is performed over all grid points at same latitude and time steps belonging to the last year of the simulation.

section 3, the disk will orbit and hence will provide time-varying insolation change across a range of latitudes, which will lead to a more uniform reduction in mean temperature, and in principle reduced regional impacts of geoengineering.

5.2 Out-of-plane Occulter

The previous first-order prediction of the impact of an occulting disk sized to offset 1.7% of solar insolation, as discussed in section 3.3, can now be more accurately determined by means of iterating using the secant method of numerical analysis on the GREB climate response to different disk sizes. The result of this process yields a 1509 km radius disk (or equivalent multiple disks) required in order to obtain a mean global surface temperature of 287.2 K, largely removing the effect of CO_2 emissions. The climate response to such a 1509 km radius disk on a double CO_2 world is shown in Figure 13. It is clear from the mean latitudinal increase of temperature in Figure 13(b), that despite the fact that the occulting disk cancels almost perfectly the global effect of climate change, strong regional effects are still clear. In fact, the mean *change* of temperature at any point over the Earth's surface, as opposed to the mean temperature, it is still of order 0.3 K.

It is understood that by placing the occulting disk on the Sun–Earth line, a shadow is cast on the Earth. However, by displacing the occulting disk from the Sun–Earth line different shading patterns can be used to reduce the impact of regional effects of climate change, as estimate by GREB. Most importantly, there exists a continuum of equilibrium solutions displaced from the classical L_1 equilibrium point that can be reached with specific combinations of occulting disk orientation and disk areal density.[22] For example, by allowing a tilt of the occulting disk of only 0.25°, and given the areal density of the minimum mass point at 2.36×10^6 km from the Earth (discussed in

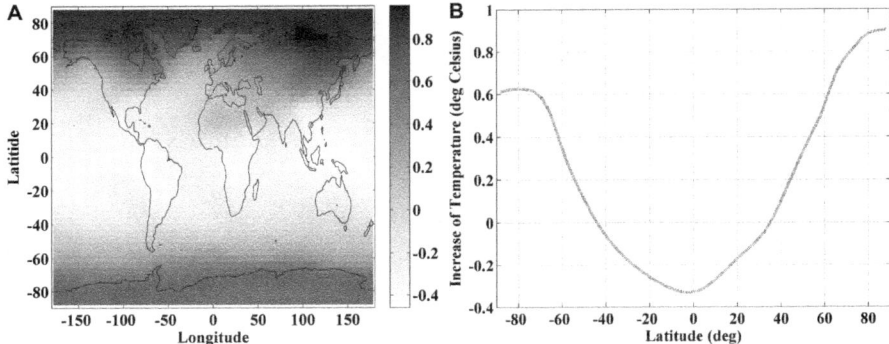

Figure 13 $2 \times CO_2$ Geoengineered world scenario as computed by the GREB model in a 50 year simulation. Scenario assumes a circular disk of 1,509 km radius placed on the Sun-Earth line at a distance of 2.36×10^6 km from the Earth. A) Average change of annual mean surface temperature. B) Change of annual mean latitudinal temperature.

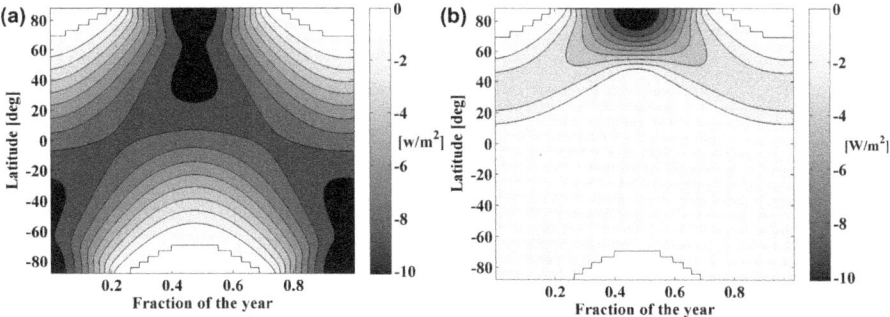

Figure 14 Impact of daily-averaged shadow over a complete year. (a) Impact of shadow cast by a 1509 km radius occulting disk on the Sun-Earth line at 2.36×10^6 km from the Earth. (b) Impact of shadow cast by a 1509 km radius occulting disk located at the same distance from Earth, but displaced by 15 000 km in the out-of-plane direction.

section 3.3), out of plane displacements of more than 10 000 km can be achieved. Note that this tilt is so small that the disk cross-sectional area, as seen from the Earth, varies only by a factor of 10^{-5} which can be considered, for the accuracy intended in this chapter, a negligible change.

Figure 14 then shows a representation of the impact of the shadow cast by a 1509 km radius disk. The shadow is represented here as the daily-averaged decrease in solar insolation as a function of latitude and fraction of year, where a fraction of 0 represents the 1st January. In particular, Figure 14(a) illustrates the impact of the shadow cast by a disk located at an equilibrium position on the Sun–Earth line and at a distance of 2.36×10^6 km from the Earth. For comparison, Figure 14(b) shows the impact of the shadow cast by the same disk, but located at an alternative position displaced by 15 000 km in the out-of-plane direction. It is clear from Figure 14 that the fraction of solar radiation intercepted by the displaced occulting disk is much smaller. Indeed, while in case A the occulting disk is intercepting approximately 1.9% of the solar insolation arriving at the Earth, in case B this falls to 0.2% due to the out-of-plane displacement. Nevertheless, one may envisage that by wasting a fraction of the total insolation reduction it is perhaps possible to find disk configurations that reduce the global mean change of temperature more uniformly across the Earth than that achieved for a disk located on the Sun–Earth line. However, it is clear that these alternative configurations with a displaced disk will require larger areas than a disk located on the Sun–Earth line.

Taking advantage of the relative linearity of the climate response to different radiative forcing patterns,[29] an optimal placement of two separate occulting disks can be sought. In particular, the objective is to find the out-of-plane displacement and size of the two occulting disks that, if stationary at out-of-plane equilibrium positions, can minimise the global mean change of temperature across the Earth, again minimising regional impacts.

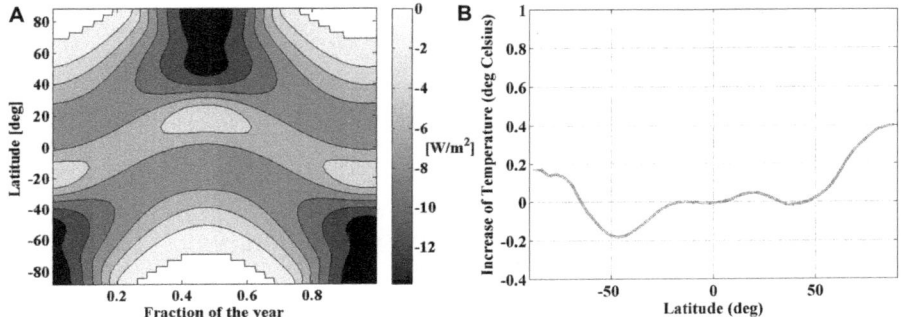

Figure 15   Geoengineering scenario with two displaced occulting disks located at
2.36×10^6 km from the Earth, but with 12,000 km symmetric out-of-
plane displacements. A) Daily-averaged latitudinal shade over a complete year. B) Climate response to a 50 year simulation with a step
change from 340 ppm to 680 ppm CO_2 concentration given the insolation change represented in A).

A space-based geoengineering scenario with two displaced occulting disks
located at a distance of 2.36×10^6 km from the Earth, but with 12 000 km
symmetric out-of-plane displacements is described in Figure 15. This shows
an important reduction of the global mean change of temperature, as can be
seen in Figure 15(b). For a single disk located on the Sun–Earth line (see
Figure 13), the mean change of temperature across the Earth was of order
0.27 K, while the scenario described in Figure 15 demonstrates a much
more uniform regional impact of the shading of the disks. However, this
scenario requires two disks of 1765 km radius, both located at a distance of
2.36×10^6 km from the Earth, with one displaced 12 000 km in the positive
out-of-plane direction and the other located symmetrically in the negative
out-of-plane direction. Hence, this scenario requires close to a 3-fold increase in shading area in order to achieve the uniform regional impact of the
disk shading. This increase in the shading area is expected however, since
the displaced positions are much less efficient at intercepting solar radiation
than locations on the Sun–Earth line.

5.3   Optimal Orbiting Disk

The problem of inefficient use of the occulting disk area can be partially
alleviated by considering, instead of stationary displaced positions, artificial
motion that allows the occulting disk to move, enabling seasonal variation of
insolation change across Earth's surface. As can be seen in Figure 15(a), the
impact of the shading of the displaced disks has a minimum impact during
the (northern hemisphere) winter months. Therefore, if the displaced disk
was to orbit, it may be possible to achieve a uniform reduction in temperature as described in Figure 15 with only one disk.

A periodic out-of-plane displacement is now sought such that the mean
change of temperature over the Earth's surface is minimised. This goal is
met with an artificial orbit such as the one shown in Figure 16, where the

Cartesian coordinates have origin at the classical L_1 Lagrange point with the x-axis along the Sun–Earth line, z normal to the Sun–Earth line and y completing the triad. Figure 16(a) represents a disk orbit with a sinusoidal motion of amplitude 14 595 km. Initial conditions and control laws can then be found that allow the required orbit to be followed by small changes in the orientation of the occulting disk. The control laws are shown in Figure 16(b) defined by the cone angle α and clock angle δ as a function of time.[23] The cone angle α defines the angle between the Sun-line and the normal to the surface of the occulting disk, while the clock angle δ is defined here as the angle between the y-axis and the projection of the normal to the surface of the occulting disk on the y–z plane.

An occulting disk of 1970 km radius, following the orbit described in Figure 16, now delivers the climate response described in Figure 17, with a

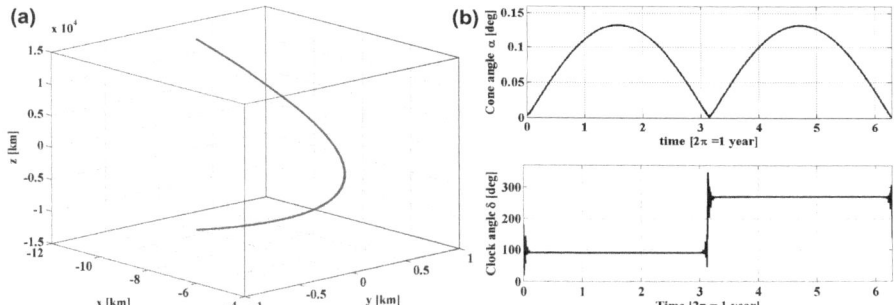

Figure 16   Artificial Geoengineering orbit of amplitude 14 595 km. (a) 1 year period motion in the Earth rotating reference frame, centred on the classical L_1 point. (b) Cone and clock angle (α, δ) control law required to generate the required orbit.

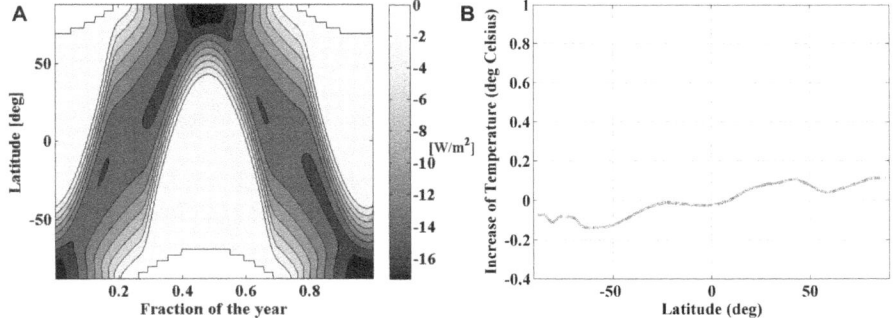

Figure 17   Climate response to artificial Geoengineering orbit with a disk of 1,970 km radius. A) Daily-averaged latitudinal shade over a complete year. B) Climate response to a 50 year simulation with a step change from 340 ppm to 680 ppm of CO_2 concentration given the insolation change generated by the disk motion and represented in A).

mean change of temperature across the Earth of order 0.06 K. The advantage of this scenario over the two symmetric disk case is clear: in this case only a 70% increase in shading area is necessary to provide a uniform insolation reduction with minim regional impacts. The motion of the disk itself should not represent a more significant engineering endeavour than that required to place and station-keeping a disk on the Sun–Earth line. Any of these options will require periodic station-keeping manoeuvres to keep the disk in place and with the correct attitude due to the instability of the classical L_1 equilibrium point.

6 Conclusions

This chapter has offered insights into the possibilities offered by space-based geoengineering using both orbiting solar reflectors and artificial dust clouds. While such space-based geoengineering schemes clearly require a leap of the imagination over current large-scale terrestrial or space engineering endeavours, the natural and human-driven variability of the Earth's climate will necessitate some form of manipulation of solar insolation in the very long-term. Again, while the scale of engineering discussed in this chapter is clearly daunting, the continuous availability of vast quantities of solar energy in space, and the active control of such energy using orbiting thin film solar reflectors may allow the possibility of large scale manipulation of the climate. Whether such possibilities are exploited in the future, both to mitigate natural and anthropogenic climate change on Earth and to unlock the resources of space, remains to be seen.

References

1. J. H. Ausubel, *De-carbonization: The Next 100 Years*, A.M. Weinberg Lecture, Oak Ridge Nat. Lab., June 2003.
2. D. W. Keith, *Ann. Rev. Energy Environ.*, 2000, **25**, 245.
3. M. I. Budyko, *Climatic Changes*, American Geophysical Union, Washington, 1977.
4. P. J. Crutzen, *Clim. Change*, 2006, **77**, 211.
5. J. Latham, *Atmos. Sci. Lett.*, 2002, **3**, 52.
6. S. Salter, G. Sortino and J. Latham, *Philos. Trans. R. Soc. London, Ser. A*, 2008, **366**, 3989.
7. R. E. Zeebe and D. Archer, *Geophys. Res. Lett.*, 2005, **32**, L09703.
8. J. Lehmann, J. Gaunt and M. Rondon, *Mitigat. Adapt. Strategies Global Change*, 2006, **11**, 403.
9. R. Angel, *Proc. Nat. Acad. Sci. U. S. A.*, 2006, **103**, 17184.
10. C. R. McInnes, *Proc. Inst. Mech. Eng. Part C: J. Mech. Eng. Sci.*, 2010, **224**, 571.
11. T. M. Lenton and N. E. Vaughn, *Atmos. Chem. Phys.*, 2009, **9**, 5539.
12. D. J. Lunt, A. Ridgwell, P. J. Valdes and A. Seale, *Geophys. Res. Lett.*, 2008, **35**, L12710.

13. W. Seifritz, *Nature*, 1989, **340**, 603.
14. J. T. Early, *J. Br. Interplanet. Soc.*, 1989, **42**, 567.
15. H. Hudson, *J. Br. Interplanet. Soc.*, 1991, **44**, 139.
16. M. Mautner, *J. Br. Interplanet. Soc.*, 1991, **44**, 135.
17. C. R. McInnes, *J. Br. Interplanet. Soc.*, 2002, **55**, 307.
18. C. R. McInnes, Planetary macro-engineering using orbiting solar reflectors, in *Macro-engineering: A Challenge for the Future,* Springer, London, 2006.
19. M. Mautner and K. Parks, in *Engineering, Construction and Operations in Space II*: Proceedings of Space 90, American Society of Civil Engineers, 1990, vol. 2.
20. P. Pearson, J. Oldson and E. Levin, *Acta Astronom.*, 2006, **58**, 44.
21. C. Struck, *J. Br. Interplanet. Soc.*, 2007, **60**, 82.
22. C. R. McInnes, A. J. C. McDonald, J. F. L. Simmons and E. W. MacDonald, *J. Guidance, Control Dynamics*, 1994, **17**, 399.
23. C. R. McInnes, *Solar Sailing: Technology, Dynamics and Mission Applications,* Springer-Verlag, London, 1999.
24. J. Shepherd, K. Caldeira, P. Cox and J. Haigh, *Geoengineering the Climate*, Report of the Royal Society working group of Geoengineering, 2009.
25. R. Bewick, J. P. Sanchez and C. R. McInnes, *Adv. Space Res.*, 2012, **49**, 1212.
26. M. Wilck and I. Mann, *Planet. Space Sci.*, 1996, **44**, 493.
27. B. Govindasamy and K. Caldeira, *Geophys. Res. Lett.*, 2000, **27**, 2141.
28. D. Dommenget and J. Floter, *Clim. Dynamics*, 2011, **37**, 2143.
29. A. B.-W. George and C. Ken, *Environ. Res. Lett.*, 2010, **5**, 034009.

Solar Radiation Management and the Governance of Hubris

RICHARD OWEN

ABSTRACT

Proposals for the intentional engineering of the Earth's climate through techniques of solar radiation management (SRM) have been accompanied by profound questions of governance. As the purpose, goals and motivations of SRM are considerations of paramount importance, governance must not only encompass risks and unintended consequences, but also intent. In this chapter, I pose two questions as these relate to SRM and governance. Firstly, should we be entertaining the thought of research or deployment of SRM and its governance, *i.e.* is SRM a legitimate object of governance, and if so under what conditions? And, linked to this, secondly, is SRM governable, particularly within democratic political systems? Arguing that SRM is a political artefact I will describe some potential problems it may present for democratic governance. I will go on to sketch a brief history of governance discussions and initiatives concerning SRM. In doing so I will observe that the boundary work of learned societies, some academics and others has attempted to legitimise SRM research as an object of governance, defining governance contours and thresholds, underpinned by normative principles. I will review some recent personal experiences of the first attempt to move from words to actions, in terms of governing a SRM research project within a framework for responsible innovation. I will finally review the results of emerging public and stakeholder dialogue exercises which reveal that while attitudes towards SRM research are nuanced and ambivalent, publics and many

Issues in Environmental Science and Technology, 38
Geoengineering of the Climate System
Edited by R.E. Hester and R.M. Harrison
© The Royal Society of Chemistry 2014
Published by the Royal Society of Chemistry, www.rsc.org

stakeholders have great antipathy, even hostility, towards SRM deployment. As research is projected through to deployment both become simultaneously framed and the legitimacy of SRM research questioned. Conditions for acceptable deployment that include the need for international agreement and governance may be perceived as being highly implausible, with concerns that SRM may prove incompatible with governance based on democratic principles, and may generate unprecedented forms of geopolitical conflict. Given these considerations I will conclude that the question of whether SRM, and its research, is a legitimate object of governance remains to be democratically decided, if indeed it ever can be.

Let us go then, you and I,
When the evening is spread out against the sky
Like a patient etherised upon a table;

T.S. Eliot

1   Introduction: Hubris, Piety and the Limits of Human Governance

In the *Historia Anglorum* (or the History of the English People) the medieval chronicler Henry of Huntington recounts the legend of how Cnut, the 11th Century King of Norway, Denmark and England, had his chair carried to the English sea shore, where he commanded the tide to halt. As the tide continued to rise 'without respect to his royal person' he leapt from the chair declaring

> '*Let all the world know that the power of kings is empty and worthless and there is no King worthy of the name save Him by whose will heaven and earth and sea obey eternal laws*'.

Cnut's actions, often misrepresented as hubris, were in fact a demonstration of piety. In his world the eternal laws of nature were beyond the will of Kings and mortal men: they were only governable by God. The medieval, deontological society of Cnut has all but disappeared, although many still believe in this divine corporation view and the limits of human governance, royal or otherwise. Others of an atheist or agnostic persuasion may also recognise such limits, instead taking the position that there are some things that are *not governable at all*: by humans, Kings or God. They may acknowledge that there are some things (*e.g.* laws of chemistry, physics and such like) that we may be able to understand and use to our advantage (and which we have indeed used to change our environment, sometimes on a spectacular scale). There are other things we may be able to predict such as volcanic eruptions, but over which we have no control. And there are still other things, such as earthquakes, that we can neither predict nor control.

Those who have ever been at sea in even the most modest of storms, or in the path of a tornado, hurricane, or tsunami, will need no reminder of the power of nature, the fragile relationship we have with it, our own vulnerability, finitude and the limits of human control. In the face of those things that are ungovernable by man, if one is not inclined to be pious then there is at least a place for humility.

There is however another constituency of thought emerging, one that has arguably evolved from our history: this, at least in the West, first asserted that it is God's will that man exploit nature for his proper ends and subsequently, through the Enlightenment,[1] Ascent of Science (and its fusion with technology) and the Industrial Revolution, grew to conceive nature as a set of laws and processes that can be observed, learned, harnessed and controlled.[2] In the spirit of a modern day Prometheus, this world view perceives the limits of human ambition and its governance as being set only by ourselves. It perceives science and innovation as an endless frontier where nothing, including nature, is beyond human understanding, use and control, if only we put our minds to it; a Baconian relationship with nature defined in terms of mastery and even domination.[1,3] It is this hubris, and aspirations to govern it, that this chapter is concerned with. I will be discussing the governance of research and (possible) deployment of techniques known collectively as solar radiation management (hereafter which I will refer to as SRM). Robock (see Chapter 7) provides a detailed technical description of such techniques and associated potential effects, which I will not repeat here. The proposition is itself rather simple (and perhaps it might be argued rather elegant) and the technologies involved might even be described as being rather mundane, albeit deployed at a grand scale[4]: by increasing albedo, or the Earths ability to reflect back a small proportion of incoming solar radiation (or insolation), we might be able to induce a cooling effect, reducing global warming, by up to several degrees and in a relatively short timescale, perhaps a matter of years or even months. Proposals to modify the weather are hardly new, dating back at least to the 1830s when American meteorologist James Pollard Espy proposed controlled forest burning as a means to stimulate rain.[5,6] The context in which discussions concerning the research and possible use of SRM techniques are currently occurring is rather different. This is one of runaway rises in atmospheric CO_2 (arguably at least to some degree of our own making), the threats of this in terms of greenhouse gas-induced climate change and the potential to exceed so called climate 'tipping points'.[7] This is compounded by our inability, or unwillingness, to curb global CO_2 emissions, and the sad realisation that even if we did, the latency of atmospheric carbon means CO_2 levels will inevitably continue to rise.[7]

At its core is the idea that the relationship between our species and our planet is reaching, or has reached, crisis point, where SRM may present the only option left,[2] or the lesser of evils (see Scott, 2012 and references within for a broader discussion of this point in the more general context of environmental ethics).[8,9] This thinking argues that we should consider, and even have a moral duty to fund and undertake, research aimed at exploring the feasibility of engineering our global climate, perhaps at a planetary scale. Of the potential SRM techniques available two have been identified as particularly promising, on grounds of potential effectiveness, technical feasibility and cost[7,10]: cloud whitening (*e.g.* by increasing the number of cloud condensation nuclei in marine stratus clouds which form over substantial portions of the oceans *e.g.* using fine sea salt particles; and stratospheric particle injection (see Salter, Chapter 6),[11] whereby sub micron particles (*e.g.* sulfate aerosols) might be deliberately injected into the stratosphere (*e.g.* at approximately 20 km altitude and at a rate of several Tg S per year, (see Robock, Chapter 7) *via* a number of potential delivery mechanisms. The latter could, it is argued, allow us to mimic the atmospheric temperature reducing effects witnessed during large scale volcanic eruptions such as Mount Pinatubu in 1991 in which a transient global temperature drop of 0.5 °C was observed over several months. Cost, it would appear, is unlikely to be a limiting factor (see Robock, Chapter 7). It is, could, as some have stated, be cheap, fast and imperfect.[12]

It is these two forms of SRM that I will focus on. There is a great deal of uncertainty and ignorance regarding the technical feasibility, impacts and risks of such techniques. However what is clear is that such forms of SRM are aimed at alleviating the symptoms of lifestyle-associated disease in *Homo anthropocenus*, rather than providing a cure. Rather than addressing our obsession with growth and consumption and its associated high carbon lifestyle (in particular in the developed world and compounded by exponential population growth, especially in the developing world) – *i.e.* rather than addressing the causes of the problem, which are both moral and political in nature – SRM would serve only to treat some of its symptoms, while offering the potential to introduce uncertain and unevenly distributed side effects. These might include regional impacts, for example on precipitation, hydrological cycles (including possibly significant effects on tropical monsoons), polar ozone, and feedback effects which could counteract or reinforce those associated with climate change itself and which would be differentially distributed. Such symptomatic treatment would not constitute a one-off course: it would require periodic, or even continuous and possibly intergenerational administration.[7] It would require a delicate balance to be struck between reduced insolation and continuing greenhouse warming, perhaps for centuries. Sudden cessation of SRM could result in rapid temperature and precipitation rises at 5 to 10 times the rates of gradual warming (see Chapter 7),[12] and the effects associated with proposed cessation would then have to be balanced against those of continued use.

Critically, it seems clear that deployment would not ameliorate some effects of greenhouse gas accumulation (*e.g.* ocean acidification) nor return us automatically to some previous, desirable or steady climatic state. It would create a *new climate*, one which might benefit some, might not benefit others, may harm others still and may also harm many if we chose, or had to stop it. It could pose serious moral issues of restitution and intergenerational justice, *i.e.* as a new climate it might deny future people choices and opportunities they might otherwise have had, with little opportunity to opt out or go back.[3] Deployment would constitute an endless experiment with nature, and the societies that inhabit it.[4] Arguably we have been unintentionally changing our climate for some time, at least since the industrial revolution, but this would be different: it would be *intended*,[10] an important moral distinction.[3] And while all forms of life modify their contexts to some degree,[1] SRM would, properly, constitute *an end of nature* where global systems would be essentially linked to human choices,[2] managed and controlled by us at an unprecedented scale, the ultimate embodiment of Han Jonas' diagnosis of the altered nature of human action, mediated by technology.[13]

What counts as SRM research may not be readily apparent. Some SRM research might for example look rather similar to other climate related research, and *vice versa*: the difference may only be the intentions of the researcher(s) (see Heyward,[14] and Boucher *et al.*,[15] for further discussion on this point).[10] We must therefore talk primarily about the *governance of intent*,[16] rather than the *post hoc* governance of unintended consequences as these relate to our environment, of which we have some experience to draw on (*e.g.* chlorofluorocarbons (CFCs) and their impact on the polar ozone layer,[17] and the intercontinental transport and impacts of persistent organic pollutants such as organohalogens: there is plenty written on these subjects). These issues of motive, purpose and intent are critically important and can become matters of great concern among the public,[18] particularly in contentious areas of new technology such as genetic modification.

I would like to suggest there are two key questions relevant to governance as this relates to the research and use of SRM to engineer the climate. The first question is primarily an ethical one, a question of *should*. Is the intention to make a new climate using SRM *the right thing to do*, and is it something we should even be thinking about researching and attempting to govern? In other words, should SRM be, on normative and ethical grounds, *an object of governance,* and if so *under what conditions*?[3] There are at least two answers to this question: we should not entertain the thought of SRM research or application and the development of processes to govern this, *i.e.* SRM and its research are not a legitimate object of governance. We might collectively decide on a moratorium, or even a ban – which somewhat ironically might necessitate governance itself, albeit narrowly framed as ensuring, preventing or deterring SRM activity (*e.g.* research, field trials, deployment), if this were indeed possible. Or we might answer yes, SRM deployment and/or research constitute a legitimate object of governance, outright or with certain conditions – which begs the question what conditions would be applied,[3,4,19] and how could we ensure these are democratically arrived at.[8,18,20]

There are no specific regulations relating to SRM.[7] There are more general conventions relating to transboundary harm caused by *e.g.* atmospheric pollutants which could potentially encompass SRM, although enforceability would be an outstanding question. An agreement to prohibit large scale geoengineering under the UN Convention on Biodiversity was for example reached in 2010, but this is not legally binding, with few compliance structures and limited remit (see Olson, 2011, p. 38).[21]

I am going to argue that even so, what might be described as forms of *de facto* governance, and in particular the *boundary work* of experts (*e.g.* through their visions and judgements) and learned societies (*e.g.* through their reports), has attempted to legitimise SRM research as an object of governance, specifying certain normative principles and thresholds.[†] This boundary work has begun to identify the contours of, and conditions for, such governance, which I will describe. In doing so it has drawn a distinction between research (and within this certain thresholds of research) and use, a moral division of labour between science and application, which may have important consequences (such as the technological lock described in ref. 22). The question of whether this attempted legitimation is *democratic* remains, I suggest, open: it is as I will go on to describe, certainly contested. Despite this, SRM research (of many different kinds it must be stressed, including to some small degree social and political) *is* underway, with normative principles being declared under which it should be conducted and practical attempts at governance being experimented with, including attempts of my own with others, that I will later describe.

The second question is linked to the first, and is a '*can*' question. It asks whether SRM can be (practically, feasibly) governable and if so, how. SRM could certainly pose significant challenges for governance (see Box 1).

Clearly if one feels SRM is not practically governable, particularly using the institutions present in and between countries based on principles of democracy, then it is hard to argue for SRM as a legitimate object of governance. In fact both questions are profoundly Cnutian in nature, in that they challenge us to ask *whether SRM is, and should be, beyond the governance of man*. We all, collectively, find ourselves seated at the edge of the shore in this regard. They are not simple questions, and embed a raft of issues, from considerations of risk and uncertainty, to the status of knowledge, to issues of equity, power, intergenerational justice, values and, not least, our relationship with and place in the natural world. Those of a pious nature, those who feel SRM lies beyond the limits of human governance, may be inclined to jump up from the chair and be done with it. Those of a more hubristic disposition may be inclined to remain seated and try to find a way to metaphorically govern the waves, or rather skies, exploring what governance of SRM research, and even application, might look like: I will certainly attempt to provide some insights into the former that are emerging from the literature. For those of you who remain unsure, I will leave you to ponder your own position for now, and in fairness there is some detail that should be described and which may inform your position (not that this is my goal).

[†]The fact that this chapter, and others in this book, may also contribute to such boundary work is not lost on me.

Box 1 A few governance challenges for SRM.
         (Adapted from Ref. 10, 4 and Robock, Chapter 7).

> ➤ How can international agreement over the 'ideal' global climate be reached?
> ➤ Who should decide, and on the basis of what criteria, where and when SRM field experiments and deployment should occur? Is it possible to come to such a decision democratically?
> ➤ Can legitimate, collective and democratic control over SRM deployment that some might seek to do unilaterally be established?
> ➤ Will SRM catalyse or require autocratic forms of governance?
> ➤ Can governance processes be developed, evolved and accommodated within existing democratic governance structures, including legal constructs, on a national and international scale?
> ➤ Could SRM lead to transferring risk to the poorest countries and the most vulnerable people?
> ➤ How would liability and compensation for adverse impacts, including on a trans-national and intergenerational scale, be handled? How would contested views concerning complex attribution of weather events to natural variation or SRM be handled and resolved?
> ➤ Can, and should, intent and motivation for SRM, which will always be plural, ever be governed, and if so how and by whom, at to what ends?

The remainder of the chapter is laid out as follows. I will first review in a rather general, and it has to be said sometimes speculative way, some features of SRM (as we know it) and how these may relate to governance. SRM has a vocabulary that includes ignorance, uncertainty, ambiguity and contingency. I hope then that these features will be treated as issues for consideration, rather than immutable facts. There is a small but growing literature in terms of SRM and its governance and no end of speculation to draw on. There is also a rich *corpus* of knowledge in terms of the social and political constitution, and governance, of technologies which serve as important heuristics, foundations and signposts for this, one of its most hubristic examples. I will then describe how SRM has been subject to various forms of *de facto* governance which have collectively attempted to legitimise its research as an object of governance, a process which has and continues to be highly contested.[4,23] In doing so I will also attempt to address the second question, *i.e.* is SRM and its research governable and if so how, including some of my own experiences. I will finally consider some illuminating, and rather ominous, recent work which has considered broader views (*e.g.* of publics) and their thoughts on SRM and its governance, which are highly germane to both questions.

2   SRM as Political Artefact

Governance of SRM can be defined in a number of ways. The Solar Radiation Management Governance Initiative (SRMGI) report that I will go on to describe refers to it as 'resources, information, expertise, and methods needed for the control of an activity, in order to advance the potential societal benefits provided by SRM, while managing associated risks'.[10] I find this to be a somewhat narrow and possibly instrumentalised framing in that the advancement of SRM and its benefits may appear implicit if we can manage risks, but it conveys the notion of a network of actors who exert influence over the direction, trajectory and conduct of SRM research and make decisions concerning its deployment. Governance can have different functions and operate at a number of different levels from regulation of many different types,[24-26] through voluntary codes of conduct (*e.g.* the European Commission Code of Conduct for Responsible Nanotechnologies Research,)[27] to governance by market choice. It can be prefixed by a number of words – innovation, political, democratic – all of which are relevant to SRM, and indeed many other emerging technologies. The word 'democratic' is an important adjective. Democracy in this regard can be considered as a 'heterogeneous set of subnational, national and supra-national practices, principles and institutions that serve to constitute citizens as part of a collectivity, able to act freely and equally, either directly or through elected representatives, in the practice of political self-determination'.[28] Such practices, principles and institutions include: political pluralism; free and fair elections; equality before the law; protection of civil liberties; freedom of speech; sovereignty of national governments; ability to get redress for harm through legal systems; a minimal level of human rights; and a functioning of civil society.

It may at first appear simpler to identify technological solutions (a 'technofix') than it is to resolve moral and political problems,[9] until one realises that technological solutions themselves can be morally and politically constituted, and morally and politically entangled. It is well known that governance, power and technologies are interlinked, that technologies are socially constructed and that they embed political dimensions.[29] I am not going to review the social constructivist literature concerning technologies here, but suffice to say that it shows that technological things are social and political, as well as technical in nature.[18,30] This social and political aspect of their being can be emergent, often in an unpredictable way, in which unintended impacts of some type *must be expected to occur*. Technologies can also be selected, socially constructed and purposed/re-purposed with the intention of producing particular economic, social and political consequences, of which dual use of technologies (*e.g.* for military and terrorist purposes) is just one obvious example. Technologies can also be made political by design: the incorporation and embedding of certain values (social and political) into design is well known.[31,32] In other words, as Langdon Winner famously stated, 'artefacts can have politics.'[29] The objective of SRM to increase planetary albedo is as much a political project as it is a technical one. But this is not a stable political artefact: it is and will be associated with instability, dynamism and plurality in terms of its framing and goals (see Robock, Chapter 7). These goals might include: addressing threats to food or water supply; environmental objectives (*e.g.* to protect vulnerable polar regions and stem the loss of Arctic sea ice, to stop sea level rise); or goals (possibly simultaneous) that are humanitarian, commercial or military (see SRMGI, Box 2.1,[10] and Bipartisan Policy Center Task Force on Climate Remediation Research).[33] They may belie a range of motivations which are unlikely to exclude those of a political and commercial nature.[10] The goals of, and motivations for SRM are far from clear, or agreed (see Robock, Chapter 7). There is considerable opportunity for SRM to become conditioned, and even stabilised by powerful economic and political interests, and not just those in opposition to carbon mitigation measures, in particular if there are considerable commercial or political gains to be made. Intent will be *interpretively flexible* (this is not uncommon for new technologies) and challenging to govern. The governance of purpose and motivations *via* Hayekian principles of revealed consent through market choice may neither be possible, nor desirable.[26]

Technologies can not only catalyse or be selected to advance or hinder particular forms of politics, but, as Winner[29] went on to describe, they may be (in)compatible with, or *require particular forms of political governance*. They can be 'unavoidably linked to particular institutionalised patterns of power and authority' (see also Joerges for further discussion).[34] It is prescient to ask what forms of political governance could SRM be (in)compatible with, or even require? The inability to reach global agreement on climate change mitigation, and in particular carbon emissions reductions, has put democratic processes under significant strain, and some might argue has constituted a failure of democratic governance.[8,28] At the very least it suggests that political institutions quite possibly lack the capacity to govern the development and deployment of SRM[10] SRM could in fact pose serious challenges for the processes and institutions of liberal democracy. The production of novel climate configurations might for example raise complex issues of justice and compensation. The natural and anthropogenic (SRM) origins for observed impacts on, for example, weather systems, wind speeds and ocean currents might easily become conflated, with cause and effect hard to attribute. Rayner *et al.* (2013) provide a hypothetic example whereby any unusual weather event (for example, something similar to the Pakistan floods of 2011) that occurred during the execution of a large scale field test might be blamed on such a test.[35] This would place strain on legal constructs of accountability, liability and compensation. It would inevitably lead to contestation, and may also cause conflict at national, regional or global scales, for example if SRM were pursued unilaterally by countries or wealthy institutions/individuals.[12] The potential for this, combined with the instability and plurality of framings and motives, may necessitate closed forms of decision making and forms of centralised, autocratic governance incompatible with the principles of democracy, in which a democratised world might ironically and tragically survive its own implications only through the dismantling of democracy itself: what Szerszynski *et al.* describe as a 'centralised, autocratic, command-and-control world-governing structure'.[28] They question the very notion that SRM is a legitimate candidate for democratic governance. The counter argument to this (and one that has been levelled at Winner's theories) is one based on contingency and the dangers of speculative ethics *e.g.* given the unpredictability of technologies, how can we be sure this autocratic constitution will be needed or indeed emerge, and surely if this is an undesirable outcome then, to echo David Collingridge's aspiration for corrigibility, can we not steward SRM towards democratic governance? Perhaps: I believe this at least to be a fair counter argument. And there are others (*e.g.* Kruger)[36] who contend that universal democracy is not a prerequisite for SRM but that engagement with representatives of countries which may be affected by it should be 'sincere, thorough and transparent'; he does not stipulate that those representatives should be democratically elected.

There are three summary points that emerge from the above discussion: firstly that SRM is likely to be a technology (or technologies) that are inherently political in the sense of being favourable to certain patterns of social relations and unfavourable to others; secondly that as a result there is 'an urgent need to make explicit the particular way in which SRM is being constituted as a technology, to interrogate the embedded assumptions and sociopolitical implications of this constitution, to question whether it might encourage forms of politics that may be incompatible with democratic governance, and to explore the specific challenges that SRM might pose to democracy itself'[28]; and finally, that the possibility that SRM has the potential to generate geo-political conflict and require (even instigate) autocratic forms of governance is a possibility not to be ignored. These are central considerations for the governance of SRM in democratic societies.

3 SRM Research and Attempts to Legitimate it as an Object of Governance

3.1 The Royal Society 2009 Report

I have so far presented SRM as an emerging technology defined by purpose(s) which may be co-opted for different goals and with different underlying motives (including political ones), and be (in)compatible with, or even require certain forms of political governance. It is a political artefact. I have described some emergent views as to why SRM may prove problematic in the context of democratic governance. I have characterised SRM as being interpretively flexible and unstable in terms of its framing and motives (see Selin, and references within for further, more general reading in this area).[37] In reality, these aspects of SRM and its governance have been only little explored to date.

It is naive to equate governance solely with regulation or legally binding conventions, of which I have already stated there are none specific for SRM. Governance can take many forms,[25,26] some of which may be as important as regulation, particularly in the context of new technologies. Of these, various forms of *de facto* governance,[38] sometimes overt, sometimes tacit, sometimes covert, are important in terms of framing technologies, and influencing their directions, trajectories and pace. In the case of SRM, the visioning and boundary work performed by key 'enactors' (who may represent a spectrum from strong advocacy to vehement detraction) – scientists, social scientists, funders, learned societies, journalists, activists for example – has been critical in terms of framing SRM and has included attempts to legitimise SRM research as an object of governance.

Of these, arguably one of the more significant pieces of boundary work was the report *Geoengineering the Climate, Science, Governance and Uncertainty* by the UK Royal Society in 2009.[7] It is evident from the title of this report that governance was a key consideration. The Foreword to reports such as these can be as important as their content. In the report Lord Rees, then President of the Royal Society, framed geoengineering as follows:

> *'nothing should divert us from the main priority of reducing global greenhouse gas emissions. But if such reductions achieve too little, too late, there will surely be pressure to consider a 'plan B' – to seek ways to counteract the climatic effects of greenhouse gas emissions by 'geoengineering'...the Royal Society aims to provide an authoritative and balanced assessment of the main geoengineering options. Far more detailed study would be needed before any method could even be seriously considered for deployment on the requisite international scale.'*[7]

The report aimed to clarify scientific and technical aspects of geoengineering, and contribute to debates on climate policy. It attempted to be inclusive in its evidence gathering, including some consultation and dialogue with the public. It is important to note that it considered geoengineering in its broadest sense, *i.e.* both carbon capture and SRM approaches, clearly distinguishing between the two. The Foreword was also clear about framing – as an option of last resort that should not serve to distract us from the priority of emissions reductions, but an option that should be properly researched. Since SRM might be the only option for limiting or reducing global temperatures rapidly it should, the report argued, be the subject of further scientific investigation in the event that such interventions become urgent and necessary.[7] This would serve to 'arm the future' with knowledge and additional options for managing the climate whilst continuing with mitigation efforts.[8] The report would be a balanced, authoritative assessment by experts which attempted to legitimise, and even authorise, SRM research and its governance, beginning to define research thresholds *e.g.* between laboratory and small field trials on one hand and large scale (*e.g.* transboundary) field trials and use on the other, which would subsequently become an important narrative. It catalysed the funding of research in the area of SRM in the UK, as I will later describe.

In a similar vein, in 2011 the US Bipartisan Policy Center's report on climate remediation recommended that the US Federal Government should embark upon a focussed and systematic research programme, arguing that if the climate system were to reach a climate tipping point and swift remedial action were needed then the US government would need to be in a position to judge whether geoengineering techniques could offer a meaningful response.[33] This research should develop capabilities and assess effectiveness and risks, to include field research as well as modelling and laboratory studies, accompanied by 'competent, prudent and legitimate governance',[33] see also U. S. Government Accountability Office.[39]

This sort of boundary work is not uncommon for emerging areas of techno-science. Technology assessment of nanotechnology in 2004 by the Office of Technology Assessment at the German Bundestag for example performed a kind of boundary work on nanofuturism. In the UK, in the same year, the Royal Society and Royal Academy of Engineering's report on nanosciences and nanotechnologies performed a similar function,[40] exorcising visions of 'nanobots' and 'grey goo' which were considered to be a 'distraction' from the real issues, focusing attention through expert analysis, and a measure of public and stakeholder deliberation, on the far less exotic, and arguably less contentious, engineered nanoparticles and thereby framing and legitimising a research agenda that largely stands to this day.

There is a rational, well established logic behind this: decisions based evidentially on knowledge (broadly constituted) and good science are the best ones. Policy should be evidence-based. But in doing so the report introduced a moral division of labour between research and application of SRM, distinguishing between the governance of small scale research and the governance of large field trials and deployment, arguing the need for the former while thinking about, or even preparing the ground for the latter. The caveat here is that research does not necessarily mean 'use' (see Morrow *et al.*),[41] for a discussion of analogies of geoengineering with medical research in terms of ethical principles and precedents in fields such as medicine where there is an ethical distinction between medical research and medical practice). This is a distinction which, as I will describe later, is not necessarily one that is generally held for SRM.

Rees' successor at the Royal Society continued this narrative, arguing that, faced with an impending *grand malum*, there is almost a moral obligation to research such techniques, in terms of feasibility, efficacy, safety and effects, even if the decision to use such techniques is the privilege of others and one, we may hope, that never has to be taken.

"One would not take a medicine that had not been rigorously tested to make sure that it worked and was safe. But, if there was a risk of disease, one would research possible treatments and, once the effects were

established, one would take the medicine if needed and appropriate.
Similarly we need controlled testing of any technologies that might be used
in the future" (Nurse, 2011, cited in Owen, 2011).[19]

If a decision has to be taken in the future concerning whether SRM presents the lesser of two evils (*i.e.* as opposed to the impacts of climate change) the argument is that such a decision should have a firm basis in good science undertake beforehand (see Gardiner[8] for further discussion on this argument).

This is echoed by those academics who posit the need to develop *capability* to do SRM in a manner that complements emission cuts, while managing the associated environmental and political risks.[12] They argue that it would be reckless to conduct the first large scale SRM tests in an emergency and that there is an immediate need for a carefully designed, incremental, transparent and international programme of SRM research, including small scale field trials (arguing it is impossible to identify and develop techniques without field testing), linked to activities that create norms and understanding for international governance of SRM.[12] Some ask whether it is indeed unethical *not* to investigate a technology that might prevent widespread dangerous impacts associated with global warming, and not provide policy makers in the near future with detailed information about the benefits and risks of various geoengineering proposals so they can inform decisions about implementation: 'only with geoengineering research will we be able to make those judgements' (see Robock, Chapter 7). If such research were blocked, only 'unrefined, untested and excessively risky' approaches would be available, constituting a 'policy train wreck'.[42]

The Royal Society recognised that the 'acceptability of geoengineering will be determined as much (if not more) by social, legal and political issues as by scientific and technical factors' and that 'there are serious and complex governance issues which need to be resolved if geoengineering is ever to become an acceptable method for moderating climate change'.[7] It saw the solution to this lying in research, development, demonstration and robust governance.

Despite advocating that geoengineering proposals should be primarily evaluated on the basis of four criteria – effectiveness, timeliness, safety and cost – it also recognised the importance of public attitudes, social acceptability and political and legal feasibility. It advocated the exploration of geoengineering governance challenges as a priority and that appropriate governance mechanisms would be needed *before* deployment of any geoengineering technology with trans-boundary implications, other than those aimed at greenhouse gas removal. It recommended research and development to investigate whether low risk methods could be made available 'if it becomes necessary to reduce the rate of warming this century'. This should include appropriate observations, the development and use of climate models, and carefully planned and executed experiments.[7] This research should be conducted in an open, transparent and internationally coordinated manner. It recommended the development and implementation of governance frameworks to *guide both research and development in the short term, and possible deployment in the longer term,* including the initiation of stakeholder engagement and a public dialogue process.[7] Any trans-boundary experiments should be subject to some form of international governance, preferably based on existing international structures.

The framing of 'properly governed research now with no presumption of use, and no deployment without international governance' has been echoed elsewhere. The American Meteorological Society (AMS) for example adopted a policy statement calling for research in July 2009, which was endorsed by the American Geophysical Union and readopted by the AMS in 2013.[43] This recommends enhanced research on the scientific and technological potential for geoengineering the climate system, including research on intended and unintended environmental responses; co-ordinated study of historical, ethical, legal, and social dimensions of geoengineering that integrates international, interdisciplinary, and intergenerational issues and perspectives and includes lessons from past efforts to modify weather and climate; and the development and analysis of policy options to promote transparency and international cooperation in exploring geoengineering options along with restrictions on reckless efforts to manipulate the climate system (see Rayner *et al.*[35] and Robock, Chapter 7, for other examples of calls for research in this vein).

3.2 Development of Normative Principles for Governing SRM Research

In the absence of regulation and other codifications of social norms, the drawing up of voluntary codes of conduct/practice for research in areas of emerging technologies and techno-science is one favoured option (see European Union, *Code of Conduct for Responsible Nanotechnologies Research,* 2008).[27] As I have described, one of the key recommendations in the Royal Society report was the development of a research governance framework, to include codes of practice for the scientific community. Rayner *et al.*,[35] and Kruger[36] describe how, shortly after publication of the report, in November 2009, two of its authors (Steve Rayner and Tim Kruger at Oxford University) initiated the development of a set of normative principles for governing geoengineering research which would subsequently become known as the 'Oxford Principles' (see Box 2).[35,36] This, Kruger describes, was initiated in response to the UK House of Commons Science and Technology Committee call for evidence into the regulation of geoengineering.

These academics and three others with expertise in social science, risk, international law and ethics prepared *Draft Principles for the Conduct of Geoengineering Research* which were submitted to the Science and Technology Committee.[44] The Members of Parliament, according to Kruger, used the Oxford Principles as a framework for questioning those who gave oral evidence to their enquiry, and stated in their report that 'while some aspects

Box 2   Normative Principles for Governing Geoengineering Research.

Oxford Principles for Governing Geoengineering Research.[35,44]

1. Geoengineering to be regulated as a public good
2. Public participation in geoengineering decision-making
3. Disclosure of geoengineering research and open publication of results
4. Independent assessment of impacts
5. Governance before deployment

Asilomar Principles for Responsible Conduct of Climate Engineering Research.[47]
The Asilomar Principles propose the need for international governance and suggest several elements important to governance:

1. Collective benefit
2. Establishing responsibility and liability
3. Open and co-operative research
4. Iterative evaluation and assessment
5. Public involvement and consent

Bipartisan Policy Centre Principles for Climate Remediation Research.[33]

1. Purpose should be to protect the public and environment from potential impacts of climate change and climate remediation technologies
2. Field deployment inappropriate at this time
3. Basis and direction of research based on independent advice from experts and government officials, informed by a robust process of public engagement
4. Transparency
5. International co-ordination
6. Ongoing assessment and adaptive management

---

of the suggested five key principles need further development, they provide a sound foundation for developing future regulation. We endorse the five key principles to guide geoengineering research'.[45] Responding to the report, the UK Government welcomed the outline set of principles.[46] Kruger goes on to describe how the principles were then presented at the US Asilomar Conference on Climate Intervention Technologies in March 2010 organised by the Climate Institute (which consciously drew on the famous 1975 Asilomar Conference on Recombinant DNA Technologies) where they subsequently formed the basis of the Asilomar Principles for Responsible Conduct of Climate Engineering Research (see Box 2),[47] see also Olson.[21]

The Oxford Principles, in the words of its authors, 'signal core societal values that must be respected if geoengineering research, and any possible deployment, is to be legitimate'. Intended to guide the collaborative development of geoengineering governance, from the earliest stages of research, to any eventual deployment they contain the principle of "governance before deployment" *i.e.* one that does not advocate eventual deployment, but indicates that any decision to deploy or not must be made in the context of a strong governance structure. Rayner *et al.*[35] frame it as a process to stimulate an open debate about what values should underpin a geogengineering governance regime, and what this could look like *i.e.* what operational features of a governance regime are desirable. With both normative and process dimensions, they are analogous to high level legal principles, not intended to direct action and being similar to the codes of conduct used by medical professions and beyond. In a similar vein, the US Bipartisan Policy Center's task force on climate remediation also developed a set of six principles for guiding research (see Box 2).

These sets of principles have some distinct commonalities. They advocate firstly that geoengineering (including SRM) should be of *collective benefit*, *regulated as a public good* and for *the protection of the public and the environment*. Explaining the principle of regulation as a public good in more detail, Rayner *et al.*[35] go on to state that since all humanity has a common interest a stable climate and the means by which this is achieved, the global climate must be managed jointly, for the benefit of all and with appropriate consideration for future generations, *i.e.* invoking the concept of (intergenerational) justice. This utilitarian view does not preclude private sector involvement in technique development or commercialisation, but they argue that SRM should be undertaken in the public interest by 'appropriate bodies at a state and/or international level' (see also Parson and Keith),[42] such that activities are not dominated by a small group (*e.g.* subset of governments or business interests): activities should be governed in a way that benefits everyone and that does not privilege certain interests (*e.g.* through the patent system) in an equitable and democratic manner. There should therefore be a presumption against exclusive control of geoengineering technologies by private individuals or corporations, with fair access to the benefits of geoengineering research. I will return to this issue presently as it proved important for governance in practice (see section 4).

Aligned to this, all three sets of principles advocate public participation in geoengineering decision making, and extend this to introduce the principle of informed consent of 'those affected by research activities', which in the case of SRM would require global agreement (see also Morrow *et al.*)[41] Inclusive deliberation is an important feature of the governance framework I and other colleagues developed for the SPICE SRM project that I will describe below. Kruger[36] draws on Stirling[20] – who in turn draws on Fiorino[48] – in terms of the rationale for this as being normative, *i.e.* it is the right thing to do, legitimising decision making and substantive, *i.e.* that it makes for better decision making through an inclusive approach (see Sykes and Macnaghten for an extended discussion).[49] The informed consent principle is drawn from (bio)medical ethics (Morrow *et al.*,)[41] and introduces an important nuance to the framing presented by Nurse in his quote above. Rayner *et al.*[35] make a salient point that the mode and extent of participation will depend on global differences in political and legal cultures, where there will be different ideas about democracy and different understandings of consent.

All three sets of principles also advocate disclosure, transparency and open publication of research results (*e.g.* through production of an open research register) and international co-ordination and co-operation. Transparency is a value that repeatedly emerges as a necessary component of any geoengineering governance framework.[50] Without transparency, Rayner *et al.* argue, an agent is effectively "kept in the dark", with the danger of exploitation on the one hand, or benign but disrespectful paternalism on the other. Disclosure and open publication support informed consent (see Dilling and Hauser for further discussion)[50] and promote integrity of the research process, trust and the preventing of a backlash against geoengineering researchers and their research (see Kruger).[36] Linked to principles of openness, transparency and participation, all three sets of principles also advocate, in the spirit of technology assessment, iterative and independent assessment of impacts (environmental, socio-economic) of research, including the mitigation of risks of lock in (see section 3.4).

There is some distinction between the three sets of principles: the Bipartisan Policy Center's principles advocate no deployment at this time; in contrast the Oxford Principles advocate governance before deployment; meanwhile The Asilomar principles seem to skirt this issue, although in the preamble they do assert the need for international governance. The Asilomar principles instead advocate the principle of establishing responsibility and liability. This latter principle, as I and others have discussed above and extensively elsewhere, is a particularly challenging goal.[51,52]

There are some obvious ambiguities and tensions inherent within the principles (for example what constitutes 'benefit to all' and what constitutes 'independence' of assessment) that the authors recognise. The principles also combine elements of the emerging field of responsible innovation (see section 4) which broadly has both normative aspects and process dimensions under conditions of uncertainty and contingency, and which itself builds on concepts of anticipatory governance,[24,53] technology assessment,[54–56] and so called 'upstream engagement' (Sykes and Macnaghten, and references within).[49] As such SRM and geoengineering is emerging as an important location for exploring in a more general way the governance of emerging technologies.

3.3 The Solar Radiation Management Governance Initiative (SRMGI)

Following publication of its report, in March 2010 the Royal Society entered into a partnership with the Environmental Defense Fund (EDF) and TWAS, the UNESCO academy of sciences for the developing world, to investigate governance issues raised by research into SRM. The partnership initially intended to produce some specific governance recommendations for SRM research, but then changed emphasis, instead aiming to provide a forum to open up and document governance discussions that drew in different perspectives, rather than producing prescriptive recommendations. It therefore intentionally did not act as a normative guide or code for governance of SRM research but represented a set of perspectives on governance (from no special governance to complete prohibition) as a platform for further discussion and debate. Its working groups focussed on the mechanics of SRM governance, international dimensions, thresholds and categories of research and goals and concerns. It did not attempt to distinguish what types of research would require what forms of governance. It instead focussed on the functions of SRM governance, what existing international treaties and institutions might be of relevance, ways of co-ordinating and delivering SRM governance, and how a phased adaptive approach to SRM research governance might proceed.

A key question for the SRMGI was whether research explicitly focussing on SRM has any characteristics that warrant particular (and possibly novel) forms of oversight *i.e.* in addition to the norms and rules of funding, research and publication of results (including policies of open access) and ethical review procedures at research institutions . As 'strategic research' the report argued that wider publics have legitimate interest in what kinds of research are being undertaken on their behalf and whether that exploration poses a risk to them, warranting public oversight and being open to global scrutiny (this is one of the normative principles described in the Oxford Principles). Since this is a novel proposition to research technologies that, if deployed, would intentionally change the living conditions of many people across many borders, SRM research, the report concluded, may warrant global (and possibly different) forms of governance: in this regard SRM research was considered a candidate for special consideration.

3.4   Thresholds and 'Differentiated Governance'

In general, international laws and conventions provide a largely permissive framework for geoengineering research activities.[7] A cautious approach which permits carefully controlled scientific research in the field of ocean fertilisation had already been adopted under the *London Convention and London Protocol*, see Box 4.3, Royal Society.[7] The SRMGI report concluded, however, that there are few international governance mechanisms available to ensure that SRM research would be transparent, safe and internationally acceptable. It also argued that a moratorium on research would be difficult, if not impossible, to enforce.[10]

Drawing a distinction between different types of SRM research, from computer modelling to global testing, the SRMGI report argued that effective governance should be based on differentiated governance arrangements for different kinds of SRM activity.[10] This was an approach that the report noted had been adopted through the 2010 decision by the *UN Convention on Biological Diversity*, signed by 193 countries, which states that 'no climate-related geoengineering activities that may affect biodiversity take place... with the exception of small scale scientific research studies that would be conducted in a controlled setting'.[57] What was small and large scale was not defined, but the principle of thresholds and differentiated governance was set. The SRMGI report went further, defining more precise categories of research for differentiated governance:

1. *'Indoors and passive observations'*: non hazardous studies with no potential environmental impacts such as modelling studies, passive observations of nature and laboratory studies (not involving hazardous materials, or involving hazardous material but appropriately contained and with no deliberate, intentional release into the environment): these were considered to be activities with negligible direct risks.
2. *'Outdoors activities'*:
   (a) small field trials (including release into the environment) of a magnitude, spatial scale and temporal duration that may lead to locally measureable environmental effects considered to be insignificant at larger scales – these were considered to be activities with negligible direct risks;
   (b) medium and large scale field trials (including release into the environment) leading to measureable and significant environmental effects, categorising medium field trials as having effects at local or regional levels, but not beyond national borders and categorising large field trials as those having global or large scale effects across borders: these were considered to be activities with potentially direct risks; and
   (c) deployment, leading to environmental effects of a sufficient magnitude and spatial scale to affect global and regional climate significantly and lasting for more than one year: these were activities with potentially direct risks.

What is immediately apparent from this is the reliance on risk as a differentiator. The report itself recognised that physical risk is not the only consideration, with 'public perception' being an important dimension, itself influenced by factors such as who is undertaking the research and for what purpose, reversibility and liability arrangements. The report did not attempt to go to the next step of identifying what governance arrangements should be assigned to each of the categories above, although the categorisation above rather implicitly draws a line between on one hand indoor activities and small scale field trials involving release into the environment with only local environmental effects, and on the other medium and large field trials/ deployment, in terms of the potential for direct risks (see also Boucher *et al.*,[15] who suggest that localised climate modification should be classified as an adaptation measure as long as there is no measurable remote environmental effects). The report was careful in its use of language:

"It seems clear that large-scale SRM interventions would pose potential risks and provoke contending views that would require effective governance, whether these interventions are undertaken as operational deployments or as large-scale research. It is less clear, and less widely agreed, that smaller-scale SRM research activities pose similar challenges that would require new governance mechanisms" (SRMGI, p. 29).[10]

The strategy of differentiated governance based on thresholds has also been recommended by a number of academics in the field (*e.g.* Cicerone[58] and Robock, Chapter 7). Robock distinguishes between research and deployment in terms of environmental impact, asserting that indoor research has different ethical issues to that conducted outside. In his view, curiosity-driven indoor research cannot and should not be regulated if it is not dangerous, but any emissions to the atmosphere should be prohibited if they are dangerous. Here indoor research is framed as ethical and necessary to provide information to policy makers in order to make informed decisions in the future, and outdoor research is unethical unless subject to governance that protects society from potential environmental damage. Parson and Keith[42] have also recommended a strategy based on 'defining thresholds, accepting oversight'. Asserting that low-risk, scientifically valuable research should be allowed to proceed and that large regulatory burdens could create incentives to mislabel the research's purpose, they identify three next steps to 'break the deadlock on governance of geoengineering research': (a) that government authority should be accepted – asserting that an approach of Polanyi-esque self regulation is unacceptable,[59] they advocate informal co-ordination by research funders and regulatory agencies, but with no new laws; (b) that a moratorium should be declared on large scale geoengineering with a possible 'large scale threshold' such that there is no detectable climate signal; and (c) that a 'small scale threshold' be defined below which research may proceed, based on existing regulations, possibly with modest new requirements and transparency.[42] Parson and Keith suggest thresholds based on a product of the area, duration and size of radiative forcing perturbation.

4 From Saying to Doing: Governing SRM Research within a Framework for Responsible Innovation

In response to one of the Royal Society's key recommendations (for government and research councils to fund a ten year programme of research), in October 2009 the UK research councils convened a workshop to scope a programme of geoengineering research aimed at allowing the UK to make an informed and intelligent assessment about the development of climate geoengineering technologies. Following this, in mid March 2010 several of the UK Research Councils, under the leadership of the Engineering and Physical Sciences Research Council (EPSRC), convened a funding 'sandpit' on the topic of geoengineering. Sandpits are an innovative funding approach in which participants (*e.g.* scientists) are encouraged to work across institutions and disciplines to develop novel project ideas over an intensive couple of days with help from mentors, and using an iterative process of real time peer review, with the intention of funding one or more projects by the end of the process. The geoengineering sandpit resulted in two projects being funded. One was a desk-based project aimed at developing an integrated assessment framework and tools for assessing geoengineering proposals, the other was a project called SPICE: Stratospheric Particle Injection for Climate Engineering. The aims of SPICE were, broadly, to investigate: (a) what types of particles could be injected into the stratosphere for the purposes of SRM and in what quantity; (b) how these particles could be deployed stratospherically; and (c) what impacts might be associated with deployment. The second objective included a proposed field trial in which a hose would be tethered to a balloon at 1 km altitude, through which small quantities of water would be pumped; the aim was to understand the dynamics and behaviour of the tethered balloon configuration in order to inform the design of a 20 km high deployment system (see Figure 1).

It was an engineering 'testbed' with no likely direct impacts and which easily fell under 'a small scale threshold below which research may proceed, based on existing regulations, possibly with modest new requirements and transparency'.[42] The testbed passed through the ethical approval processes at the universities concerned with little or no comment.

Given the known wider dimensions (and sensitivities) of SRM outlined in the Royal Society report and elsewhere it was proposed during the sandpit that the funds for the proposed field trial be made available subject to an independent 'stage gate' review. Stage gating is an established mechanism used in innovation management (particularly in new product development) in which investments in the innovation process are phased (or staged), with decision 'gates' where decisions are made to progress, stop, refine, redefine *etc.*, usually on the basis of technical feasibility, market potential and risk.[60] Having decided that this governance approach would be used, it then became necessary to define firstly what criteria would be used at the decision gate to support a decision to allow the field trial to go ahead (or not) and secondly who would make the decision. A meeting was convened in the late Autumn of 2010 to consider this. Representatives of the research councils, SPICE team (scientists and engineers) and social scientists (including at least one of the authors of the Royal Society report and Oxford Principles) struggled to develop a consensus on this, indeed I recall a lively discussion. At the end of the meeting I spoke with the EPSRC representatives about drawing together some criteria based on the discussions and further insights from the concept of responsible innovation I and others had been thinking about, notably as this applies to the activities of research funders.[61] Box 3 summarises these criteria, with more details provided by Stilgoe *et al.*, (2013).[62]

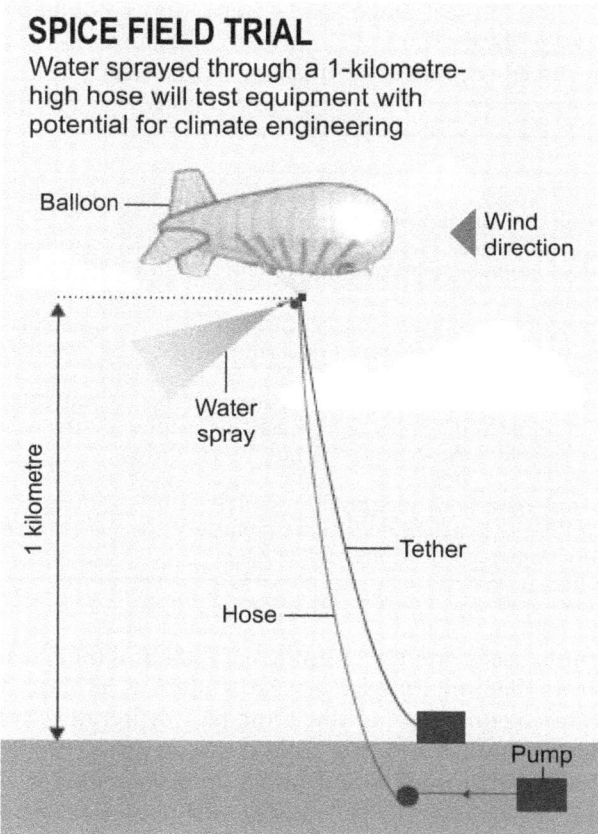

Figure 1 SPICE Testbed.
(From Macnaghten and Owen, 2011).[65]

They involved consideration of direct risks, safety and regulatory compliance (*e.g.* for flying tethered balloons) associated with the testbed itself (criteria 1 and 2), for which the SPICE team were asked to submit a risk register and statement of regulatory compliance. The third criterion required the SPICE team to reflect on the project's framing and communication, asking them to develop a communication plan to allow dissemination about the nature and purpose of the testbed (this also refers to the normative principle of transparency in Box 2). This plan was to be informed by dialogue with stakeholders. Criterion 4 asked the SPICE team to anticipate, reflect on and describe the envisaged applications of their research and the impacts (intended or otherwise) these applications may have, and embed mechanisms to review these as more information became available in the future (given the inevitable uncertainty associated with the research). It asked them to broaden their visions of application and impact, to think through other pathways to other impacts, to contextualise their work within a review of the

Box 3 SPICE SRM project stage gate criteria and responsible innovation framework dimensions.
(Reproduced with kind permission of Elsevier).[62]

Criteria	R I Dimensions
1. The testbed deployment is safe, the principal risks have been identified and managed, and are deemed acceptable	Reflexivity
2. The testbed deployment is compliant with relevant regulations	Reflexivity
3. The framing of the project (nature, purpose) for external communication is clear and advice regarding this has been obtained	Reflexivity, Inclusive deliberation
4. Future potential application(s) and associated impact(s) have been described and mechanisms put in place to review these as significant information emerges	Anticipation, Reflexivity
5. Mechanisms have been identified to understand wider public and stakeholder views regarding these envisaged applications and impacts	Inclusive deliberation, Reflexivity

known or potential risks and uncertainties of SRM and the questions (social, political, ethical) that might arise as the testbed is projected through to deployment. This again refers to the principles of iterative evaluation and assessment described in Box 2. The final criterion asked the SPICE team to identify mechanisms to understand public and stakeholder views around the project and its envisaged applications and potential impacts, and the understandings, assumptions, uncertainties, framings and commitments associated with these. This builds on the principles of public participation and engagement in Box 2. This was in part informed by a series of micro-deliberative public forums,[63] which I will describe in more detail in the next section. These criteria were aligned to a framework for responsible innovation I had been developing with others and which is in turn based on the need for research and innovation to be *anticipatory, reflexive, inclusively deliberative* and (ultimately) *responsive* (to such anticipation, reflexivity and deliberation) in terms of its direction and trajectory, in such a way that innovation and its underlying purposes, motivations and impacts are opened up,[51,62,64] empowering a measure of social agency in technological choice.[20] These dimensions are mapped on to the criteria in Box 3.

Having defined the criteria around a framework for responsible innovation and agreed how the SPICE team, working with others, might respond, it was then necessary to define how these responses would be evaluated and by whom. The stage gate panel that was convened to undertake this task did so in June 2011 and comprised two social scientists, an atmospheric scientist, an engineer with expertise in high altitude balloons, and an advisor to an environmental NGO,' observed by members of the research councils and myself. observed by members of the research councils and myself. The panel was charged with providing a recommendation for each of the five criteria (pass, pass pending further information, fail) to the research councils who would make the final decision concerning the future of the testbed.[65]

The panel decided, after considerable discussion, rebuttal and debate, that the first two criteria, concerning safety, risks and regulatory compliance associated *with the testbed itself*, were convincingly passed: there appeared to be little concern about the direct risks, environmental or otherwise, of releasing a bath tub full of water over an airfield in an unpopulated location. This, as I will very shortly describe, was not to be the issue with SPICE.

The other three criteria would only be passed pending further work and provision of further information. There were particular concerns regarding the need for a communications strategy informed by stakeholder engagement and underpinned by substantive public dialogue, more anticipation and reflection concerning the testbed and projection through to deployment (in terms of different, plausible pathways through to application and the impacts and implications the tested and envisaged applications may have – social, ethical, environmental, intended and unintended), and, finally, substantive engagement with stakeholders concerning the project and its intended application(s). The governance process asked the SPICE scientists and research funders to consider the wider (*e.g.* social) dimensions of a technoscience 'in the making', one in which established role responsibilities (of both scientists and funders) were challenged and broadened,[66,67] and one in which the very premise of the independent republic of science and its role responsibilities were questioned.[59] This was a draining but important experience for many concerned. It was also clear that it would require resourcing and support, (for example the commissioning of the public engagement work described by Pidgeon *et al.*). Ultimately it raised questions about the way the project had been set up largely as one investigating technical feasibility and environmental impacts, but not the social, ethical and political dimensions I have described in previous sections of this chapter. Amongst these was the question of whether the project should have been funded at all. It is very important to note in this regard that it was made clear by the research councils at the beginning of the stage gate meeting that the ethical question of whether the SPICE project should have been funded was not for discussion: whether SPICE should have been made an object of governance (using the framework we had devised or otherwise) was not for debate. This, many (including myself) feel was a distinct limitation and I remain of the firm belief that the process would have been a far better, and more legitimate one, had the dimensions of responsible innovation been in place for use by the research councils at the original 2009 workshop and 2010 sandpit in which decisions to fund geoengineering research, and if so of what type and in what way, were made.

On September 26th 2011, following a meeting with myself, the Stage Gate Panel Chair (Phil Macnaghten) and members of the SPICE team, the research councils decided to postpone the testbed until the pending actions had been addressed, with the intention of convening the stage gate panel again to review this later that year (see Appendix 1). On that very same day the research councils received a letter, copied to the then UK Secretary of State for Energy and Climate Change and signed by more than 50 NGOs, demanding that the project be cancelled. The NGOs saw the testbed as symbolic, sending the wrong signal to the international community, deflecting political and scientific attention from the need to curb greenhouse gas emissions.[68] There was grave concern that its 'sole purpose is to engineer the hardware that would later allow chemicals to be injected into the stratosphere to reflect sunlight' as 'a dangerous distraction from the real need: immediate and deep emissions cuts'. This would 'condemn future generations to continue a high-risk, planetary-scale technological intervention that is also likely to increase the risk of climate-related international conflict'.

With mounting interest in the media and beyond, the SPICE team began to address the outstanding stage gate criteria. It was as part of the subsequent discussions that the projects principal investigator became aware of the existence of a patent application for the balloon-tethered hose delivery system,[69] submitted by one of the sandpit mentors just prior to the sandpit itself and including two of the SPICE project scientists as named inventors. Although an internal review conducted later by EPSRC found no evidence that research council policies on vested/conflict of interest had been broken, it was clear that the patent posed a significant issue for the project in terms of the nature of at least some of the participants motivations, as well demonstrating a lack of disclosure, which was hardly in the spirit of the Oxford Principles. In May 2012, after discussions between the research councils, the SPICE team and myself, the principal investigator of the SPICE team decided to cancel the testbed (see Appendix 1), instigating a more formal process of stakeholder engagement (see section 5) which at the time of writing is ongoing.

5   A Social Licence to Operate?

"Any response to a global problem might be rejected as illegitimate and unacceptable if the majority of the world's population played little role in ... approving the response" (SRMGI, p. 25).[10]

There has been only limited stakeholder and public engagement concerning SRM. It should be noted that both the Royal Society and SRMGI reports both included consultation with stakeholders and the public. Since then there have been a few academic studies which provide some interesting insights concerning perceptions and framings of SRM and its governance. Stilgoe *et al.* describe some preliminary results of stakeholder engagement around the SPICE project,[23] which highlights the fact that questions of purpose and motivation were of paramount importance. Aligned to this, governance thresholds for research and deployment proposed by the SRMGI[10] and Parsons and Keith[42] were deeply contested. A primary reason for this is what the authors describe as 'the imaginary made real'. While stakeholders recognised that the SPICE testbed would not itself pose any direct risks it was perceived as a *symbolic act,* a potential signifier of intent. There were concerns that research may generate its own momentum and create a constituency in favour of deployment and/or that the UK might be preparing to proceed down a different strategy to carbon mitigation and adaptation using a very high risk technological approach, representing a slippery slope.[3,8]

This was not about risks but about purpose and motivation: the patenting issue that surfaced as part of questioning following the stage gate process brought these concerns into sharp relief. It called into question the legitimacy of a differentiated governance strategy based on thresholds as described in section 3. Fundamentally, SRM technology was perceived as being inherently entangled with politics (see section 2), *irrespective of the type of research done.* SRM was perceived not just as a technology, but as a political artefact.

Familiarity with SRM amongst the public is as yet low,[2] with seemingly little increase in awareness over the last five years. Pidgeon *et al.*[63] describe the results of a series of focus groups undertaken in response to recommendations from the SPICE stage gate panel that used an invited micro-deliberation methodology to understand framings of SRM and the SPICE project. This revealed that almost all participants were willing in principle to allow the testbed to proceed, but that very few were comfortable with the idea of deployment. Questions that arose included those of testbed safety and direct risks, as well as more general questions that demonstrated *projection* by participants of the research through development to deployment, with questions concerning the knowledge that the testbed might provide and its utility.

The participants felt that SRM could only provide a stop gap response to climate change, *i.e.* 'buying time', with concerns about the perceived naturalness of SRM interventions *i.e.* that SRM was perceived as interfering with natural processes (this was investigated in more detail by Corner *et al.*, 2013 who found that 'messing with nature' was a dominant narrative common to the public engagement exercises they undertook, but that this constituted a subtle set of discourses).[2] Pidgeon *et al.* report that SRM was also perceived as contributing to a 'disassociation of human kind with the physical world',[63] where SRM may be thought of as being a product of a misguided world view.[9] There were also questions about governance and specifically how SRM would be regulated and communicated. An international system aimed at enabling *a global debate concerning SRM* was seen as important. I will return to this point in the final section of this chapter.

Overall, the perceptions, associations, and interpretations of SRM deployment were negative. However, this did not automatically inhibit support for the testbed when this was framed as a *strictly limited research activity.* Participants were reluctant to rule out the SPICE field trial on condition this would be undertaken as a limited science and engineering test, but at the same time they exhibited discomfort concerning what might happen if the trial went ahead. Ambivalence towards the testbed *simultaneously* translated into concerns and opposition about deployment.

5.1   Conditionality and Implausibility

Conditionality was a key observation made by Macnaghten and Szersynski who also used a deliberative focus group methodology to engage publics with SRM.[4] These authors focused explicitly on the lived future and perceptions publics had concerning the kind of world that SRM could possibly bring into being. Thematic analysis of the engagements highlighted that SRM might be publically acceptable only under very specific and highly contingent conditions (see Box 4), conditions that by and large were seen as highly implausible in terms of their potential to be met (see also Jamieson),[3] and with perceptions that SRM was an unnatural intervention (see Corner *et al.*, on this

Box 4   Perceived conditions and plausibilities for SRM.
            (Adapted from Ref. 4)

1. *Scientific robustness.* There is confidence in the science of climate change as a reliable basis for policy. Only if people believe in the ability and authority of climate science to predict with confidence can policies aimed at climate remediation gain traction. Low plausibility: this confidence was rarely held by participants.

2. *Accurate research foreseeability.* Confidence in the ability of research to anticipate reliably the side effects of SRM in advance of deployment. Low plausibility: there was little belief in the capacity of science to identify side effects reliably in advance. Perceptions of messing with nature were seen as inevitably leading to nasty surprises. We would be 'living the global experiment' which will become part of the human condition.

3. *Condition of the ability of research to demonstrate efficacy.* Participants registered considerable doubt about technical feasibility of SRM. Only on deployment could efficacy really be ascertained.

4. *Condition of good intent and effective governance.* Confidence of the motivation of SRM as being complementary to adaptation and mitigation; confidence that SRM will be used exclusively by governments with the motivation to counteract anthropogenic climate change. Low plausibility: good intentions could never be guaranteed, being potentially open to 'dual use', used to further national, regional or commercial interests at odds with the purpose of counteracting climate change.

5. *Condition of democracy.* Confidence in the capacity of existing political systems to accommodate SRM. Low plausibility: global governance intensely difficult to achieve within democratic political arrangements, with current omens (*e.g.* lack of consensus on mitigating climate change) being poor and with SRM only being governable under autocratic governance arrangements.

point),[2] one that would constitute a short term fix that increases the likelihood of geopolitical conflict and presents major threats to democratic governance.

The discussions were nuanced and not polarised, but even those participants who started from a position of conditional acceptance grew to perceive the conditions for successful and acceptable deployment as being unfeasible and implausible, *i.e.* the more people learned about SRM technology the more sceptical they became. Since effects were perceived by some to be knowable only on deployment there was scepticism of even limited research into SRM. The authors questioned whether principles of regulation of geoengineering as a public good and public participation outlined in the Oxford Principles were attainable, arguing that upon deployment SRM could only be controlled centrally and on a planetary scale, with little opportunity for opt out.

6 Conclusions: Governing a New End of History?

In this chapter I have described some emerging proposals to engineer the Earth's climate through solar radiation management, and discussed aspects of governance as this relates to both research and deployment. It is clear from this discussion that SRM presents significant governance issues. SRM is a political artefact, a type of post-normal technoscience,[70] which makes it is a far from straightforward object of governance.[4] It is also apparent that there is distinct political unease with the notion of deployment. Where the views of publics and stakeholders have been sought, these have also highlighted great concern with, and often opposition to, the possible deployment of SRM. Many scientists frame their research as objectively informing a decision that they hope will never have to be taken. Almost everyone seems to agree that if this unpalatable decision has to be made, then there must first be international agreement and robust mechanisms of international governance in place. It is also clear that the tiered governance strategy suggested by learned societies, some governments and some academics, which distinguishes between desk-based and laboratory research, and small scale field trials on one hand, and large scale field trials and deployment on the other, is a contested one. In this regard, differentiated governance that is based mostly on the potential for direct risks is wholly insufficient: while risk and uncertainty are undoubtedly important to many, issues of purpose, motivation and intent of research, development and use are key. Historically, technological governance has struggled with intent. Creating governance thresholds for research and deployment has neglected a fundamental issue: that research and projection through to application operate as simultaneous and (socially, politically, ethically) entangled frames. Here research, even of any kind, may be symbolic, a signifier of intent, the beginning of a slippery road, with concerns of moral hazard and lock-in. For some stakeholders this imaginary made real generates concerns and even hostility towards SRM research. The little public engagement that has been undertaken suggests a lack of awareness of SRM, and ambivalence towards SRM research, which leads to significant concerns when people understand more about the proposed technology and project through from the research to application. The projection, entanglement and simultaneous framing is crucial to how we think about SRM and its governance: questions about the research *cannot be disaggregated* from questions about what the research could lead to and why the research is being undertaken, precisely because, unlike other techno-scientific umbrella terms such as 'synthetic biology' or 'nanotechnology' where purpose and application may not be immediately clear, SRM is defined by its purpose, and the plurality of goals, motivations and framings that lie beneath. Research, development and application become deeply entangled and cannot be arbitrarily or artificially separated. By doing so a strategy of differentiated governance is in danger of ignoring the core, ethical questions so central to SRM.

Conditionality (*e.g.* upon there being robust mechanisms of research governance, upon there being international agreement and governance before deployment) is a key feature of governance discourses relating to SRM and its research. Rather ominously, early findings seem to indicate that there is a sense of deep implausibility that such conditions could ever be met, particularly on an international scale. Concerns about the development and use of SRM in the absence of such conditions being met are profound, and include the potential for geopolitical conflict, challenges for democracy and democratic governance, and the potential to generate autocratic forms of governance. There is a common sense that we will be living a global, social and political experiment that will redefine our relationship with nature, uncertainty and the human condition: an experiment that many have concerns will be either workable, or desirable.

And so I return finally to the questions I posed at the beginning of this chapter. Firstly is SRM a legitimate object of governance? I believe that despite the boundary work I have described which has attempted to legitimise SRM research this is a question that remains outstanding. The success of a technological fix will depend on how it is framed,[18] and who defines the criteria for success:[9] these criteria, or conditions, must be democratically and equitably defined.[3,19] It might be argued that the boundary work undertaken to date could risk creating 'high entry barriers against legitimate positions that cannot express themselves in terms of the dominant discourse' where 'normative assumptions have not been subjected to general debate'.[18] Since SRM is primarily a political artefact where, as I have suggested, research and deployment may be perceived in simultaneous frames, as awareness becomes greater I suggest this issue of legitimacy will become increasingly contested: in fact the history of emerging technologies seems to predict this, and few have been on this hubristic scale. Addressing this question in a democratic, inclusive and substantive way is, I believe, an imperative. Morrow *et al.*,[41] drawing on principles in medical research ethics, describe this in terms of a principle of respect. Here norms for conducting SRM research are located in a *prior* discussion about whether research should be conducted at all, and if so under what conditions. It involves the securing of the global public's consent, which Morrow *et al.* assert should be voiced through government representatives before empirical research begins.[41] But assuring such 'consent by proxy' begs their key question to be answered: what representative bodies if any have authority, and legitimacy, to consent to SRM research on behalf of global publics?[3]

Secondly, and linked to the first question, is SRM practically, feasibly governable? It seems to me that we can develop open and transparent forms of SRM research governance that ensure such research is anticipatory and reflexive to its possible impacts, goals, motivations, commitments and that it is inclusively deliberative, inviting perspectives, seeking questions and ensuring that SRM research and innovation are responsive in turn. We can strive to prevent path dependency and lock in.[22] We can strive to procure 'socially robust knowledge',[71] to ensure research does not lead unreflexively to development.[3] We might even underpin these with normative principles and codes of practice. We should certainly seek to open up SRM research and ensure there is social agency in the choices and directions it takes,[20] to make it more publically accountable.[18] However, it is clear that this will require international agreement, strong institutions, and (fundamentally) a distinct culture change when this comes to science, innovation and its governance more generally: this cannot be guaranteed.

It is also clear that many find anything other than extremely limited, contained types of research deeply concerning, that even research of this limited kind is also problematic for some (possibly more than some) and that the idea of deployment is unacceptable to most without conditions that may well be implausible, and even impossible to meet. Of these conditions many remain deeply sceptical that SRM deployment can be internationally agreed upon or internationally governed, and that deployment, and indeed even research, could pose significant issues for democracy and generate conflict. It is hard to conclude anything other than the fact that SRM may well be ungovernable without very significant changes to how we govern society itself. Such grand political and social experiments have been attempted before in our history, with mixed results. In this regard proposals to research and deploy SRM are, in effect, proposals for a new end of history: one that few want and one many are sceptical can be governed. I am inclined to think that this is a social and political experiment that we should embark upon not with hubris, but with a profound sense of humility.

Postscript

As I finished writing this chapter the International Panel on Climate Change published its *5ᵗʰ Assessment Report*.[72] Within this the IPCC makes explicit reference to the potential for SRM geoengineering, if realisable, to substantially offset a global temperature rise, but notes that limited evidence precludes a comprehensive assessment of SRM and its impact on the climate system. It goes on to state that SRM methods will carry side effects and long-term consequences on a global scale.

Appendix 1   Transcripts of Engineering and Physical Sciences Research Council (EPSRC) Announcements regarding postponement and cancellation of the SPICE testbed.

Update on the SPICE Project (September 29th 2011) http://www.epsrc.ac.uk/newsevents/news/2011/Pages/spiceupdate.aspx (last accessed 7/2/14)

Stratospheric Particle Injection for Climate Engineering (SPICE) is an EPSRC, NERC and STFC-funded project that includes a work package on assessing the feasibility of injecting particles into the stratosphere from a tethered balloon for the purposes of solar radiation management.

EPSRC has taken the decision to delay the experiment planned in October, to allow time for more engagement with stakeholders. We have adopted a responsible innovation approach with this project – as part of our commitment to responsible development – and our decision to pause the testbed experiment reflects the advice that we have received from our advisory panel following a stage gate.

The technology test would have involved pumping water to a height of 1 km through a suspended hose, held aloft by a helium-filled balloon. This would allow the engineers to study how the hose and balloon behave over time in a variety of weather conditions.

SPICE Project Update (May 22nd 2012) http://www.epsrc.ac.uk/news events/news/2012/Pages/spiceprojectupdate.aspx (last accessed 7/2/14)

Stratospheric Particle Injection for Climate Engineering (SPICE) is an EPSRC, NERC and STFC-funded project that is investigating the feasibility of injecting particles into the stratosphere for the purposes of solar radiation management, *i.e.* reflecting a small percentage of the sun's light and, or heat back into space.

This involves considering different types and quantities of particles and where they could, hypothetically, be injected into the atmosphere to effectively and safely manage the climate system. It is also looking into how particles might be delivered and the likely impacts on the climate and environment.

The SPICE project includes a work package to examine the viability of using a tethered balloon and hose mechanism as a delivery method to inject particles. The work package that contains this testbed element accounts for approximately £500 000 of a £1.6 million project grant.

The SPICE project team and the research councils have chosen to follow a responsible innovation approach to the project. Responsible innovation encourages approaches that can be used early on in the innovation process to promote the responsible emergence of novel technologies in society and the identification of their wider impacts and associated risks.

The responsible innovation approach for this project included a stage gate. This is where a panel of external experts considered the progress of the project against a number of criteria, such as checking that mechanisms have been identified to understand wider public and stakeholder views on the envisaged applications and impacts.

Following the stage gate meeting, the panel advised the research councils and the SPICE team that further work on stakeholder engagement and the social and ethical implications was required. In addition, EPSRC, acting on advice from the panel, decided to delay the planned testbed experiment which would have used a tethered balloon and hose to disperse water at a height of 1 km, until this further stakeholder engagement had been undertaken.

EPSRC has provided additional funding for expert researchers to carry out this work on stakeholder engagement which includes discussion of issues around the commercialisation of geoengineering research; this is in progress and will continue.

As a result of the stage gate and the responsible innovation approach, the SPICE team was also encouraged to explore issues connected to the potential future use of geoengineering technologies. Intellectual property and need for governance in the field of geoengineering became, and continue to be, matters that concern them.

Given these issues and the existence of a patent application for an invention to deliver particles *via* a tethered balloon system, the SPICE team has decided not to conduct the 1 km testbed experiment. We received formal confirmation of this from the team on May 22, 2012. This decision is accepted by EPSRC, NERC and STFC.

The SPICE team is committed to putting all the results arising from the SPICE project into the public domain, without delay and according to normal academic practice. The SPICE team and EPSRC have agreed that this should be a condition of the grant. The results arising from SPICE will not be patent-protected.

EPSRC and the SPICE team support the Oxford Principles on Regulation of Geoengineering, which were endorsed by the House of Commons Science and Technology Committee's report on *The Regulation of Geoengineering* (March 2010).

Acknowledgments

The catalyst for this chapter stems from my involvement in a recent paper with Bron Szerszynski, Phil Macnaghten, Matt Kearnes and Jack Stilgoe in which we began to think about the political dimensions of SRM and its potential implications for democratic governance.[28]

References

1. L. White Jr, The historical roots of our ecological crisis, *Science*, 1967, **155**(3767), 1203–1207.

2. A. Corner, K. Parkhill, N. Pidgeon and N. Vaughan, Messing with nature? Exploring public perceptions of geoengineering in the UK, *Global Environ. Change*, 2013, **23**(5), 938–947.

3. D. Jamieson, Ethics and Intentional climate change, *Clim. Change*, 1996, **33**, 323–336.

4. P. Macnaghten and B. Szerszynski, Living the global social experiment: an analysis of public discourse on solar radiation management and its implications for governance, *Global Environ. Change*, 2013, **23**(2), 465–474.

5. D. W. Keith, Geoengineering the climate: history and prospect, *Ann. Rev. Energy Environ.*, 2000, **25**, 245–284.

6. J. R. Fleming, *Fixing the Sky: The Checkered History of Weather and Climate Control*, Columbia University Press, New York, 2010.

7. *Geoengineering the Climate: Science, Governance and Uncertainty*, The Royal Society, 2009; royalsociety.org/uploadedFiles/Royal_Society_Content/policy/publications/2009/8693.pdf (last accessed 4/10/2013).

8. S. M. Gardiner, Is "arming the future" with geoengineering really the lesser evil? Some doubts about the ethics of intentionally manipulating the climate system, in *Climate Ethics: Essential Readings*, ed. S. M. Gardiner, S. Caney, D. Jamieson and H. Shue, Oxford University Press, London, 2010.

9. D. Scott, Geoengineering and environmental ethics, *Nature Educat. Knowledge*, 2012, **3**(10), 10.

10. SRMGI, *Solar Radiation Management: The Governance of Research*, The Royal Society/EDF/TWAS, London, 2011; www.srmgi.org/files/2012/01/DES2391_SRMGI-report_web_11112.pdf (last accessed 4/10/2013).

11. P. J. Crutzen, Albedo enhancement by stratospheric sulfur injections: a contribution to resolve a policy dilemma?, *Clim. Change*, 2006, **77**, 211–220.

12. D. W. Keith, E. Parsons and M. Morgan, Research on global sun block needed now, *Nature*, 2010, **463**, 426–427.

13. H. Jonas, *The Imperative of Responsibility*, University of Chicago Press, Chicago, 1984.

14. C. Heyward, Situating and abandoning geoengineering: a typology of five responses to dangerous climate change, *Political Sci. Politics*, 2013, **46**, 23–27.

15. O. Boucher, P. M. Forster, N. Gruber, M. Ha-Duong, M. G. Lawrence, T. M. Lenton, A. Maas and E. Vaughan, Rethinking climate engineering categorization in the context of climate change mitigation and adaptation, *WIREs Clim. Change*, 2014, **5**, 23–25.

16. J. Stilgoe, A question of intent, *Nature Clim. Change*, 2011, **1**, 325–326.

17. European Environment Agency (EEA), *Late Lessons from Early Warnings: The Precautionary Principle 1896–2000*, Office for Official Publications of the European Communities, Luxemburg, 2001. www.eea.europa.eu/publications/environmental_issue_report_2001_22 (last accessed 4/10/2013).

18. S. Jasanoff, Technologies of humility: citizen participation in governing science, *Minerva*, 2003, **41**, 223–244.
19. R. Owen, Legitimate conditions for climate engineering, *Environ. Sci. Technol.*, 2011, **45**, 9116–9117.
20. A. Stirling, "Opening up" and "closing down": power, participation, and pluralism in the social appraisal of technology, *Sci. Technol. Human Values*, 2008, **33**, 262–294.
21. R. Olson, *Geoengineering for Decision Makers,* Science and Technology Innovation Program, Woodrow Wilson International Center for Scholars, Washington, DC, 2011, pp. 39; www.wilsoncenter.org/sites/default/files/Geoengineering_for_Decision_Makers_0.pdf (last accessed 7/10/2013).
22. D. Collingridge, *The Social Control of Technology,* Francis Pinter (Publishers) Ltd, London, 1980, pp. 200.
23. J. Stilgoe, M. Watson and K. Kuo, From Bio to Geo: implications from public engagement with new technologies for geoengineering research governance, *PLos Bio.*, 2013, **11**, e1001707.
24. R. Karinen and D. H. Guston, Towards anticipatory governance. The experience with nanotechnology, in *Governing Future Technologies. Nanotechnology and the Rise of an Assessment Regime,* ed. M. Kaiser, Springer, Dordrecht, Heidelberg, London and New York, 2010.
25. E. Fisher and A. Rip, *Responsible innovation: multi-level dynamics and soft intervention practices Responsible Innovation*, ed. R. Owen, J. Bessant and M. Heintz, John Wiley, London, 2013, ch. 9, pp. 165–184.
26. R. G. Lee and J. Petts, Adaptive governance for responsible innovation, in *Responsible Innovation*, ed. R. Owen, J. Bessant and M. Heintz, John Wiley, London, 2013, ch. 8, pp. 143–164.
27. European Commission, Code *of Conduct for Responsible Nanotechnologies Research*, 2008; ec.europa.eu/research/science-society/document_library/pdf_06/nanocode-apr09_en.pdf (last accessed 4/10/2013).
28. B. Szerszynski, M. Kearnes, P. Macnaghten, R. Owen and J. Stilgoe, Why solar radiation management geoengineering and democracy won't mix, *Environ. Planning A*, 2013, **45**(12), 2809–2816.
29. L. Winner, Do artifacts have politics?, *Daedalus*, 1980, **109**(1), 121–136; reprinted in *The Social Shaping of Technology*, ed. D. A. MacKenzie and J. Wajcman, Open University Press, London, 2nd edn, 1999.
30. W. E. Bijker, T. P. Hughes and T. J. Pinch, *The Social Construction of Technological Systems: New Directions in the Sociology and History of Technology,* MIT Press, Cambridge, MA, 1987.
31. M. J. van den Hoven, Value sensitive design and responsible innovation, in *Responsible Innovation*, ed. R. Owen, J. Bessant and M. Heintz, John Wiley, London, 2013, ch. 4, pp. 75–84.
32. M. J. van den Hoven, G. J. C. Lokhorst and I. van de Poel, Engineering and the problem of moral overload, *Sci. Eng. Ethics*, 2012, **18**(1), 1–13.
33. Bipartisan Policy Center Task Force on Climate Remediation Research, *Geoengineering: A National Strategic Plan for Research on the Potential Effectiveness, Feasibility, and Consequences of Climate Remediation*

Technologies, 2011; bipartisanpolicy.org/library/report/task-force-climate-remediation-research (last accessed 23/9/13).

34. B. Joerges, Do politics have artefacts?, *Social Studies Sci.*, 1999, **29**(2), 411.
35. S. Rayner, C. Heyward, T. Kruger, N. Pidgeon, C. Redgwell and J. Savulescu, The Oxford Principles, *Clim. Change*, 2013, **121**, 499–512.
36. T. Kruger, *A Commentary on the Oxford Principles: Geoengineering Our Climate Working Paper and Opinion Article Series*, 2013; wp.me/p2zsRk-76 (last accessed 4/10/2013).
37. C. Selin, Expectations and the emergence of nanotechnology, *Sci. Technol. Human Values*, 2007, **32**, 196–220.
38. M. Kearnes and A. Rip, The emerging governance landscape of nano-technology, in *Jenseits von Regulierung: Zum Politischen Umgang mit der Nanotechnologie*, ed. S. Gammel, A. Lösch and A. Nordmann, Akademische Verlagsgesellschaft, Berlin, 2009, pp. 97–121.
39. GAO, *Climate Change: A Coordinated Strategy could Focus Federal Geoengineering Research and Inform Governance Efforts*, GAO Report 10–903, 2010; hwww.gao.gov/products/GAO-10-903 (last accessed 4/10/2013).
40. Royal Society and Royal Academy of Engineers, *Nanoscience and Nano-technologies: Opportunities and Uncertainties*, 2004; www.nanotec.org.uk/finalReport.htm (last accessed 4/10/2013).
41. D. R. Morrow, R. E. Kopp and M. Oppenheimer, Toward ethical norms and institutions for climate engineering research, *Environ. Res. Lett.*, 2009, **4**, 045106.
42. E. A. Parson and D. W. Keith, End the deadlock on governance of geoengineering research, *Science*, 2013, **339**, 1278–1279.
43. American Meteorological Society, *Geoengineering the Climate System*; www.ametsoc.org/policy/2013geoengineeringclimate_amsstatement.html (last accessed 6/11/2013).
44. S. Rayner, C. Redgwell, J. Sauvulescu, N. Pidgeon and T. Kruger, *Draft Principles for the Conduct of Geoengineering Research (the 'Oxford Principles')*, reproduced in House of Commons Science and Technology Committee, The Regulation of Geoengineering, Fifth Report of Session 2009–2010, HC221, pp. 29–30; www.publications.parliament.uk/pa/cm200910/cmselect/cmsctech/221/221.pdf (last accessed 4/10/2013).
45. UK House of Commons Select Committee on Science and Engineering, *The Regulation of Geoengineering*, The Stationery Office, London, 2010; www.publications.parliament.uk/pa/cm200910/cmselect/cmsctech/221/221.pdf (last accessed 4/10/2013).
46. UK Government, *Government Response to the House of Commons Science and Technology Committee 5th Report of Session 2009-10: The Regulation of Geoengineering*, The Stationary Office, London, 2010.
47. Asilomar Scientific Organizing Committee (ASOC), *The Asilomar Conference Recommendations on Principles for Research into Climate Engineering Techniques*, Climate Institute, Washington DC, 2010, 20006; www.climate.org/PDF/AsilomarConferenceReport.pdf (last accessed 4/10/2013).

48. D. Fiorino, Environmental risk and democratic process: a critical review, *Columbia J. Environ. Law*, 1989, **14**, 501–547.
49. K. Sykes and P. M. Macnaghten, Responsible innovation: opening up dialogue and debate, in *Responsible Innovation*, ed. R. Owen, J. Bessant and M. Heintz, John Wiley, London, 2013, ch. 5, pp. 85–108.
50. L. Dilling and R. Hauser, Governing geoengineering research: why, when and how?, *Clim. Change*, 2013, **121**, 553–565.
51. R. Owen, J. Stilgoe, P. M. Macnaghten, E. Fisher, M. Gorman and D. H. Guston, A framework for responsible innovation, in *Responsible Innovation*, ed. R. Owen, J. Bessant and M. Heintz, John Wiley, London, 2013, ch. 2, pp. 27–50.
52. A. Grinbaum and C. Groves, What is "responsible" about responsible innovation? Understanding the ethical issues, in *Responsible Innovation, ed. R. Owen, J. Bessant and M. Heintz,* John Wiley, London, 2013, ch. 7, pp. 119–142.
53. D. Barben, E. Fisher, C. Selin and D. Guston, Anticipatory governance of nano-technology: foresight, engagement, and integration, in *The Handbook of Science and Technology Studies*, ed. E. Hackett, M. Lynch and J. Wajcman, MIT Press, Cambridge, MA, 3rd edn, 2008, pp. 979–1000.
54. A. Rip, T. Misa and J. Schot, *Managing Technology in Society: The Approach of Constructive Technology Assessment,* Thomson, London, 1995.
55. J. Schot and A. Rip, The past and future of constructive technology assessment, *Technol. Forecasting Social Change*, 1996, **5**, 251–268.
56. D. H. Guston and D. Sarewitz, Real-time technology assessment, *Technol. Soc.*, 2002, **24**(1), 93–109.
57. Convention on Biological Diversity (CBD), *Climate-related Geoengineering and Biodiversity: Technical and Regulatory Matters on Geoengineering in Relation to the CBD*, 2010; www.cbd.int/climate/geoengineering (last accessed 4/10/2013).
58. R. J. Cicerone, Geoengineering: encouraging research and overseeing implementation, *Clim. Change*, 2006, **77**(3), 221–226.
59. M. Polanyi, The republic of science: its political and economic theory, *Minerva*, 1962, **1**, 54–74.
60. R. Cooper, Stage-gate systems: a new tool for managing new products, *Business Horizons*, 1990, **33**, 44–54.
61. R. Owen and N. Goldberg, Responsible innovation: a pilot study with the UK Engineering and Physical Sciences Research Council, *Risk Analysis*, 2010, **30**, 1699–1707.
62. J. Stilgoe, R. Owen and P. M. Macnaghten, Developing a framework for responsible innovation, *Res. Policy*, 2013, **42**(3), 1568–1580.
63. N. Pidgeon, K. Parkhill, A. Corner and N. Vaughan, Deliberating stratospheric aerosols for climate geoengineering and the SPICE project, *Nature Clim. Change*, 2013, **3**, 451–457.
64. R. Owen, P. M. Macnaghten and J. Stilgoe, Responsible research and innovation: from science in society to science for society, with society, *Sci. Public Policy*, 2012, **39**(6), 751–760.

65. P. Macnaghten and R. Owen, Good governance for geoengineering, *Nature*, 2011, **479**, 293.

66. H. E. Douglas, The moral responsibilities of scientists (tensions between autonomy and responsibility), *Am. Philos. Q.*, 2003, **40**, 59–68.

67. C. Mitcham, Co-responsibility for research integrity, *Sci. Eng. Ethics*, 2003, **9**, 273–290.

68. Hands Off Mother Earth (HOME), *SPICE Opposition Letter*, 2011; www.handsoffmotherearth.org/hose-experiment/spice-opposition-letter (last accessed 7/11/2013).

69. P. Davidson, H. Hunt and C. Burgoyne, Atmospheric delivery system, *Br. Pat. Application*, GB 2476518, 2011; worldwide.espacenet.com/publicationDetails/biblio?CC = GB&NR = 2476518 (last accessed 7/11/2013).

70. S. O. Funtowicz and J. R. Ravetz, Three types of risk assessment and the emergence of post normal science, In *Social Theories of Risk*, ed. S. Krimsky and D. Golding, Praeger, New York, 1992, 251–273.

71. H. Nowotny, P. Scott and M. Gibbons, *Re-Thinking Science: Knowledge and the Public in an Age of Uncertainty,* Polity, Cambridge, 2001, 166–178.

72. http://www.climate2013.org/images/report/WG1AR5_SPM_FINAL.pdf (last accessed 7/2/14).

Subject Index